D0553801

Radar Vulnerability to Jamming

For a complete listing of the *Artech House Radar Library*,
turn to the back of this book....

Radar Vulnerability to Jamming

Robert N. Lothes
Michael B. Szymanski
Richard G. Wiley

Artech House
Boston • London

Library of Congress Cataloging-in-Publication Data

Lothes, Robert N.
Radar vulnerability to jamming / Robert N. Lothes, Michael B. Szymanski, and Richard G. Wiley.
p. cm.
Includes bibliographical references and index.
ISBN 0-89006-388-5
1. Radar--Military applications. 2. Radar--Interference.
3. Electronic countermeasures. I. Szymanski, Michael B.
II. Wiley, Richard G. III. Title.
UG612.L68 1990 90-42992
623'.043--dc20 CIP

British Library Cataloguing in Publication Data

Lothes, Robert N.
Radar vulnerability to jamming.
1. Radar equipment. Interference
I. Title II. Szymanski, Michael III. Wiley, Richard G.
621.3848

ISBN 0-89006-388-5

685 Canton Street
Norwood, MA 02062

International Standard Book Number: 0-89006-388-5
Library of Congress Catalog Card Number: 90-42992

10 9 8 7 6 5 4 3 2 1

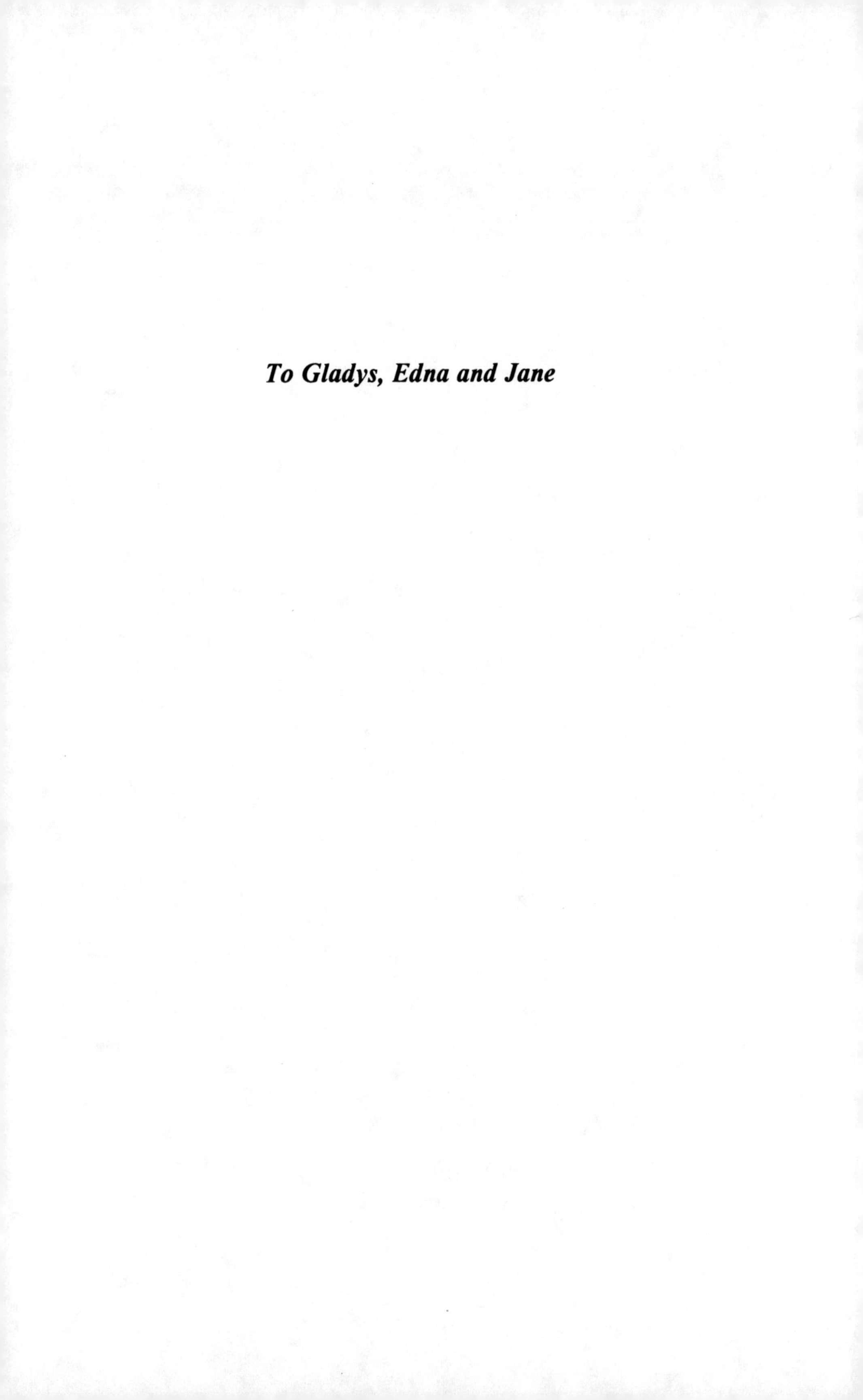

To Gladys, Edna and Jane

CONTENTS

PREFACE

This book grew out of an investigation of the vulnerability of radar to jamming done in connection with the development of computer software designed to simulate radar circuits and their response to jamming on a single pulse basis. The next step required an examination of how jamming affects closed loop subsystems within the radar over time periods which are long when compared to radar pulses or pulse repetition intervals. This book considers this "next step" and proves that using personal computers to simulate closed loop systems is an efficient method for investigating such problems.

The authors are grateful for the assistance provided by Research Associates of Syracuse, Inc (RAS); and especially for the use of computers and software and for the expert preparation of the manuscript by Mary Chamberlain.

Chapter 1
INTRODUCTION AND REVIEW OF ECM FUNDAMENTALS

1.1 SCOPE

This book deals with the effect on radars of active *electronic countermeasures* (ECM)—that is, jamming by the deliberate emission of electromagnetic radiation. The book does not deal with passive forms of ECM such as chaff and reflecting decoys. As Figure 1.1 indicates, ECM is an *electronic warfare* (EW) activity. As in other categories of warfare, the advent of new weapons stimulates the development of counter weapons and counter tactics.

The years since World War II have seen continuous improvement in radars, particularly in the areas of *radio-frequency* (RF) sources, antennas, waveform design, and signal processing. Brookner [1] extrapolates the expected course of development into the beginning of the next century. Radar development has been driven by the need for improved performance in a benign (nonjamming) environment, where the obstacles are weather, ground and sea clutter, and system noise. Development has also been driven by the need to combat ECM through the incorporation of *electronic counter-countermeasures* (ECCM) capabilities. The ECM-ECCM struggle is a continuous one, each party reacting to the advances of the other and often attempting to anticipate advances so as not to be caught unaware.

Facilitating the intelligent employment of both ECM and ECCM is the activity referred to as *electronic support measures* (ESM). ESM gear includes surveillance receivers capable of gathering radar threat information needed for the selection of jamming gear and the programming of programmable jammers to suit the mission. ESM equipment also includes radar warning receivers for alerting air crews to imminent attack by radar-directed weapons and, in some cases, for the automatic initiation of the appropriate ECM to counter a threatening radar. Although this book is mainly concerned with the understanding and evaluation of

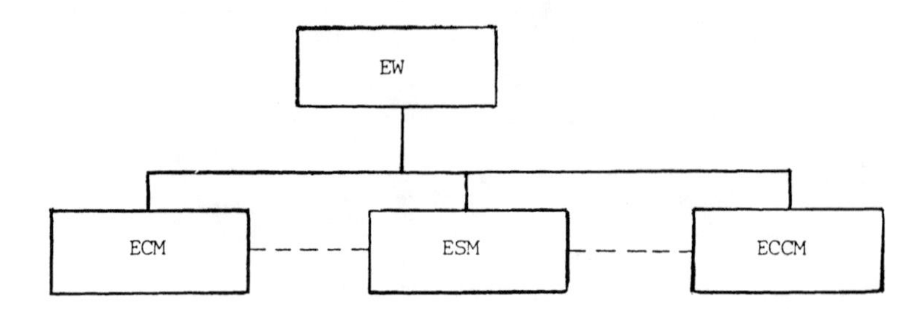

Figure 1.1 Electronic warfare activities.

various forms of jamming, we occasionally refer to the ESM outputs needed to make the jamming effective and to ESM outputs that can aid the ECCM response.

1.2 BACKGROUND

The birth of radar-related EW occurred in World War II in the air warfare over Europe and Britain. Each side sought to counteract the alerting and locating capabilities of the opponent's radars and electromagnetic navigation capabilities. The first radar-directed weapons (antiaircraft artillery) appeared during that war. In the years that followed, radar was put to extensive use for the guidance of smart weapons. Such weapons present a fearsome threat because of their long reach and high accuracy.

To the military mission planner, a quantitative estimate of jamming effectiveness is vital to the assessment of risks and the probability of success. To the team carrying out the mission, such an estimate is a matter of life and death. To the opposing force, which is dependent on its radars for defense, a quantitative estimate of the jamming threat is equally crucial. Quantitative measures are often difficult to obtain, especially when the jammer-radar interaction cannot be modeled in terms of linear differential equations. When this is the case, some form of simulation is the only realistic approach.

Wherever possible, jamming requirements are expressed as the *jamming-to-signal power ratio (J/S)* needed to produce a specified effect on radar performance. An especially useful J/S is the required ratio of jamming-to-signal power densities (watts per unit area) incident on the radar antenna, because with this ratio in hand, and with the mission geometry known, it is a simple matter to solve for the *effective radiated power* (ERP) required of the jammer. Of course, in sidelobe jamming, the radar antenna sidelobe level in the direction of the jammer must be known.

In this chapter we develop equations for several jamming scenarios, beginning with the simplest possible cases in which atmospheric absorption is negligible and all the jammer power falls within the victim receiver passband and within the receiver range gate. We then indicate the modifications needed to account for atmospheric path loss, multipath effects, and losses due to jamming power falling outside the passband or range gate. We then note that signal processing further along the receiver path may favor the signal over the jamming, thereby reducing the effective J/S at the point where the signal is utilized for target detection or tracking.

Subsequent chapters deal with specific forms of jamming and their effects on radar functions. Wherever possible, estimates of the J/S levels required to produce specific degradations in radar function are developed. The radar models used for this purpose are generic and stripped to their essentials, but we believe that the results are correct as to order of magnitude and that they point the way to more refined modeling and analysis.

1.3 RADAR-TARGET-JAMMER GEOMETRIES AND JAMMING EQUATIONS

The scenarios that follow involve an attack aircraft operating against surface installations. Whether the latter are ground-based or ship-based is immaterial. In the first two scenarios, the locations of the radar and the jammer can be interchanged without affecting the jamming equations. Moreover, the surface-based entity can be moved to an airborne platform, producing an air-to-air situation. Except for terrain-masking problems and ground-clutter considerations, the scenario can be converted to a surface-to-surface situation. The third scenario involves an attack aircraft employing a surface-based transponder beacon as a reference-point marker. Such beacons can serve as way-point markers for navigation, as markers of rendezvous locations, and, when carried by a forward air controller, as an offset aim point for the delivery of munitions against targets visible to the forward air controller but not to the air crew.

Escort or Self-Protection Jammer (SPJ)

In Figure 1.2, A is an attack aircraft and J is an ECM aircraft. The mission involves penetration into enemy airspace where the penetrator is subject to attack by surface-based, radar-directed weapons (artillery or missiles). Clearly, the jamming equations will be unchanged if the jammer is moved from the ECM aircraft (escort jamming situation) to the attack aircraft (self-protection jamming situation), for the two aircraft are considered to be flying near one another.[1]

[1]Note that if the attack aircraft and jamming aircraft ranges are comparable, but their angular separation is sufficient to place the jammer in the radar antenna's sidelobes, the jammer's impact is reduced by the sidelobe/main-lobe gain ratio.

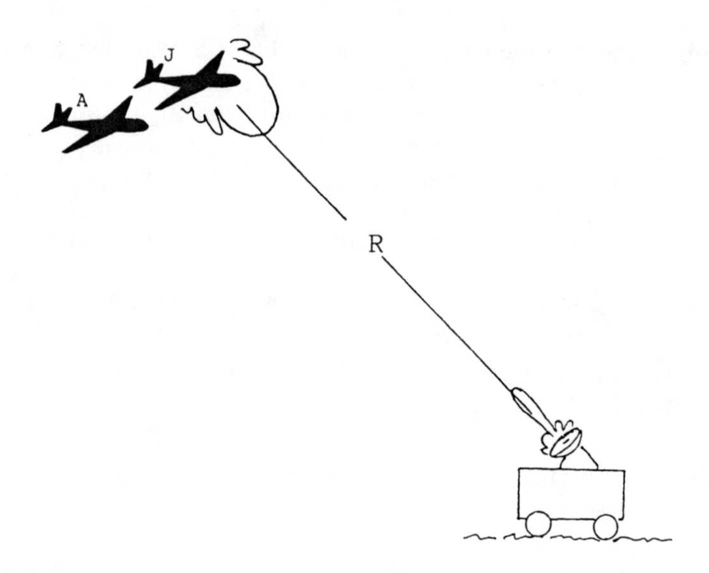

Figure 1.2 Self-protection or escort jammer scenario.

The radar emits power P_R. If it were emitted isotropically over 4π steradians (sr), the power density (watts per unit area) at range R would be $P_R/4\pi R^2$, but because the radar antenna has a gain G_R, the power density in the main-lobe region is $P_R G_R/4\pi R^2$. The quantity $P_R G_R$ is the radar's ERP. The target, with radar cross section σ_T, intercepts some of the incident radar power and backscatters it to the radar. The scattering of incident energy is not isotropic, but σ_T is so defined as to yield the actual level of backscattered power if the intercepted power (σ_T times the incident power density) is considered to be scattered isotropically over 4π sr. Thus, the power density W_S of the signal returned to the radar is

$$W_S = \left(\frac{\mathrm{ERP}_R}{4\pi R^2}\right)\left(\frac{\sigma_T}{4\pi R^2}\right)$$

where ERP_R is the effective radiated power of the radar, and R is the radar range to the target. The jammer's radiated power is P_J, and its antenna gain in the direction of the radar is G_J, so its ERP is

$$\mathrm{ERP}_J = P_J G_J$$

The jamming power density at the radar is

$$W_J = \frac{\text{ERP}_J}{4\pi R^2}$$

The radar antenna intercepts a signal power $W_S A_e$ and a jamming power $W_J A_e$, where A_e is the effective aperture area of the radar antenna. The effective aperture area is related to the gain by

$$A_e = \frac{G_R \lambda^2}{4\pi}$$

where λ is the wavelength (the value of A_e is, in general, less than the actual projected area of the antenna aperture). The signal power into the radar antenna is, therefore,

$$S = W_S A_e = \frac{\text{ERP}_R \sigma_T}{(4\pi)^2 R^4} A_e$$

and the jamming power is

$$J = W_J A_e = \frac{\text{ERP}_J}{4\pi R^2} A_e$$

When both signal and jamming impinge on the same capture area A_e, J/S can be viewed either as the power-density ratio W_J/W_S incident on the radar antenna or as the power ratio out of the antenna into the radar receiver:

$$J/S = \left(\frac{\text{ERP}_J}{\text{ERP}_R}\right)\left(\frac{4\pi R^2}{\sigma_T}\right) \tag{1.1}$$

Equation (1.1) is the jamming equation for self-protection or escort jamming in its most succinct form. The ERPs pertain to actual radiated powers and take into account transmission line losses between transmitter and antenna as well as dissipative losses in the antenna. When the angular separation between the two aircraft is sufficient to place the jammer in the radar antenna's sidelobes, (1.1) must be multiplied by the radar antenna's sidelobe/main-lobe gain ratio, G_{RJ}/G_R (see standoff jammer discussion). The jammer, striking over a one-way path, clearly has a great advantage over the signal acquired by round-trip backscatter. The magnitude of the second factor of (1.1) can be enormous, being the ratio of the area of a sphere of radius R to the radar cross section σ_T (σ_T may be on the order of a square meter).

Stand-off Jammer

Figure 1.3 describes the geometry of a jamming scenario in which the penetrating aircraft depends on the protection of a *stand-off jammer* (SOJ) aboard an ECM aircraft. This aircraft, because it can afford to be heavier and slower than an attack aircraft, is likely to carry a higher-power jamming transmitter and a larger, higher-gain antenna than is feasible for an attack aircraft. The ECM aircraft's relatively high ERP_J enables it to perform its job from a safe stand-off range, R_J. Because high gain implies narrow beamwidth [2, Section 9.1], there is a limit to the usable gain; the beam must be broad enough to ensure blanketing of the region in which the enemy's radar-directed air defenses lie, yet allow for inadvertent errors in aiming the jammer antenna.

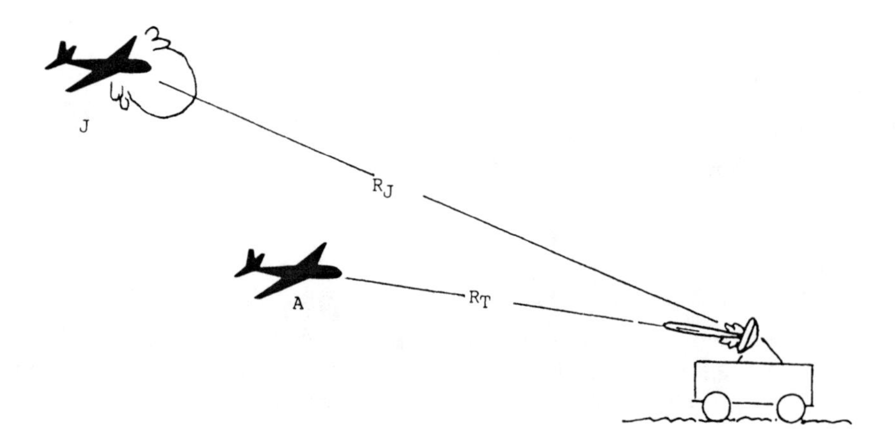

Figure 1.3 Stand-off jamming scenario.

The jamming aircraft may orbit an elongated racetrack course, the long axis of which is normal to the *line-of-sight* (LOS) to the area targeted for jamming, and transmit from one of two antennas aimed abeam the aircraft (i.e., an antenna is available on either side). With two or more jamming aircraft simultaneously on orbit, one can cover for the other as the latter executes the turn at either end of the racetrack. The SOJ needs high ERP to overcome the spreading loss of a large R_J and to enable it to inject jamming power through the radar antenna's sidelobes. As Figure 1.3 indicates, the LOS from the SOJ aircraft to the radar does not, in general, coincide with the LOS from the attack aircraft to the radar. The radar antenna gain in the direction of the target (attack aircraft) is G_R, and the gain in the direction of the SOJ aircraft is G_{RJ}. The radar antenna's effective capture areas for the two directions are, respectively,

$$A_{eT} = \frac{G_R \lambda^2}{4\pi} \qquad \text{and} \qquad A_{eJ} = \frac{G_{RJ} \lambda^2}{4\pi}$$

Now we cannot claim, as in the self-protection and escort jamming cases, that J/S into the radar receiver is equal to the ratio of the incident power densities. Instead, the values of S and J are given, respectively, by

$$S = \frac{\text{ERP}_R \sigma_T}{(4\pi)^2 R_T^4} A_{eT}$$

and

$$J = \frac{\text{ERP}_J}{4\pi R_J^2} A_{eJ}$$

where R_T and R_J are the target and jammer ranges, as indicated in Figure 1.3. The resultant J/S is, therefore,

$$\begin{aligned} J/S &= \left(\frac{\text{ERP}_J}{\text{ERP}_R}\right)\left(\frac{R_T^2}{R_J^2}\right)\left(\frac{4\pi R_T^2}{\sigma_T}\right)\left(\frac{A_{eJ}}{A_{eT}}\right) \\ &= \left(\frac{\text{ERP}_J}{\text{ERP}_R}\right)\left(\frac{R_T^2}{R_J^2}\right)\left(\frac{4\pi R_T^2}{\sigma_T}\right)\left(\frac{G_{RJ}}{G_R}\right) \end{aligned} \tag{1.2}$$

As in the previous case, the one-way *versus* two-way path advantage appears as the $4\pi R_T^2/\sigma_T$ factor. When, as in Figure 1.3, the jammer is obliged to jam via the radar antenna's sidelobes, it suffers a disadvantage equal to the radar's main-lobe/sidelobe gain ratio, which we shall call SLR:

$$\text{SLR} = \frac{G_R}{G_{RJ}}$$

In this case, the jamming equation becomes

$$J/S = \left(\frac{\text{ERP}_J}{\text{ERP}_R}\right)\left(\frac{R_T^2}{R_J^2}\right)\left(\frac{4\pi R_T^2}{\sigma_T}\right)\left(\frac{1}{\text{SLR}}\right) \tag{1.3}$$

The stand-off range disadvantage, represented by the factor R_T^2/R_J^2, amounts to an increasing penalty as the attack aircraft approaches closer and closer to the radar.

Radar Beacon Jamming (Ground-Based Beacon)

Figure 1.4 represents a scenario far less common than the preceding two, yet it merits consideration because it represents an operating mode of certain military airborne radars. B is a transponder beacon, A is an attack aircraft, and J is a jammer with the aim of jamming the uplink (the reply link). A more familiar beacon application uses a beacon as an airborne transponder for *air traffic control* (ATC) or for *identification friend or foe* (IFF) [3, Chapter 38]. The radar beacon mode is also called, especially in British literature [4, Chapter 6], *secondary surveillance radar* (SSR), as distinguished from primary (target-echo based) radar.

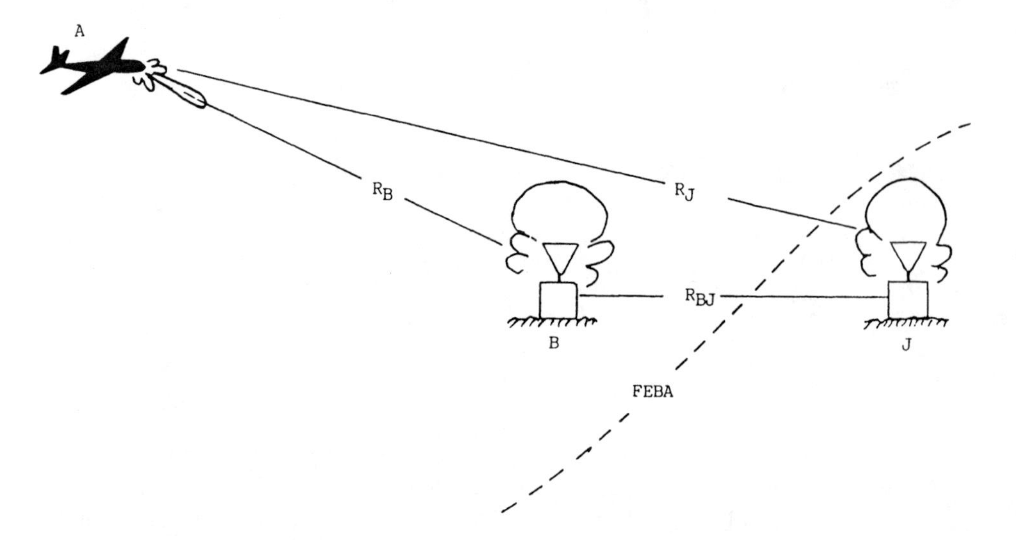

Figure 1.4 Radar beacon scenario (uplink jamming case).

In Figure 1.4, the beacon, placed by a forward air controller, marks a reference point from which the controller can designate targets across the *forward edge of the battle area* (FEBA). The attack aircraft, using the beacon as a reference, directs its attack at the designated target, for which the coordinates relative to the beacon are radioed to the aircraft by the forward air controller. Thus, targets invisible to the air crew can be pinpointed, with errors dependent only on the accuracy with which the controller performs the designation.

Each airborne radar pulse serves as a beacon interrogation pulse that elicits a beacon reply at an ERP level of ERP_B. The ERP of the jammer is ERP_J. Thus, into the radar receiver:

$$J/S = \left(\frac{ERP_J}{ERP_B}\right)\left(\frac{R_B^2}{R_J^2}\right)\left(\frac{G_{RJ}}{G_R}\right) \tag{1.4}$$

where G_{RJ} and G_R are, respectively, the radar antenna gains in the jammer and beacon directions. The J/S of interest is the one that prevails when the radar antenna main lobe is on the beacon, so G_R is the main-lobe gain. The beacon reply frequency is offset from the radar interrogation frequency, thereby permitting clutter-free beacon reception. The jammer must, of course, emit at the reply frequency. The beacon receiver threshold[2] is such that replies occur only when the radar's main lobe sweeps across the beacon with main-lobe gain G_R. If the jammer is sufficiently close to the beacon, both jammer and beacon will lie within the radar antenna main lobe, producing the ratio:

$$J/S = \left(\frac{ERP_J}{ERP_B}\right)\left(\frac{R_B^2}{R_J^2}\right) \tag{1.5}$$

If there is sufficient angular separation between beacon and jammer, the jammer will lie in the radar antenna's sidelobe region during the beacon response. If, as earlier, we let SLR represent the main-lobe/sidelobe gain ratio:

$$SLR = \frac{G_R}{G_{RJ}}$$

the sidelobe jamming equation becomes

$$J/S = \left(\frac{ERP_J}{ERP_B}\right)\left(\frac{R_B^2}{R_J^2}\right)\left(\frac{1}{SLR}\right) \tag{1.6}$$

Clearly, these jamming equations are also applicable to the jamming of any point-to-point communication link, for the beacon and the radar constitute the transmitting and receiving terminals of an elemental communication channel. The airborne radar antenna's beam is narrow (typically a few degrees), but the beacon antenna's pattern is broad to accommodate aircraft arriving from anywhere within a wide sector. Likewise, the jammer antenna's pattern must be broad unless it is

[2]The threshold may be controlled by *automatic gain control* (AGC).

steered on the basis of aircraft direction information, obtained perhaps from a nearby surveillance radar.

Evidently, main-lobe jamming occurs when the aircraft is near an extension of the beacon-jammer baseline, or when the aircraft range is very large relative to the baseline length. To a good approximation, the main-lobe jamming region can be shown to lie outside a circle that is tangent to the baseline at its midpoint. The smaller the radar antenna beamwidth, β, the larger is the main-lobe jamming circle. In Figure 1.5(a), the aircraft is approaching at angle ϕ relative to the normal to the beacon-jammer baseline, which has length R_{BJ}. If the angle α, subtended by the baseline, is less than $\beta/2$, main-lobe jamming occurs. Therefore, $\alpha = \beta/2$ defines the boundary of the main-lobe jamming region. Because $\beta/2$ is small (a degree or so), the angle subtended by the baseline when the aircraft is on this boundary can be approximated as

$$\alpha \approx \tan \alpha \approx \frac{R_{BJ} \cos \phi}{R}$$

Then, setting $\alpha = \beta/2$, we can solve for the normalized range:

$$\rho = \frac{R}{R_{BJ}} = \frac{2}{\beta} \cos \phi \tag{1.7}$$

As Figure 1.5(b) indicates, (1.7) is the polar equation of a circle of radius $1/\beta$ tangent to the baseline at its midpoint. The area outside the circle is the main-lobe jamming region where (1.5) applies. We can show that the contours of constant J/S for sidelobe jamming within the circle described by (1.7) are another family of circles (we are modeling the entire sidelobe region as a uniform-gain region at a level 1/SLR below the main lobe). The law of cosines applied to Figure 1.6(a) yields

$$R_J^2 = R_B^2 + R_{BJ}^2 - 2R_B R_{BJ} \cos \theta$$

which can be manipulated into the form:

$$r_B^2 - 2r_B\left(\frac{1}{1-k^2}\right) \cos \theta + \frac{1}{1-k^2} = 0 \tag{1.8}$$

where $r_B = R_B/R_{BJ}$ is the normalized beacon range, and $k = R_J/R_B$ is the jammer-beacon range ratio. Figure 1.6(b) is included to show that (1.8) describes a circle. Applying the law of cosines to Figure 1.6(b) yields

$$r_B^2 - 2r_B x_1 \cos \theta + x_1^2 - \rho^2 = 0 \tag{1.9}$$

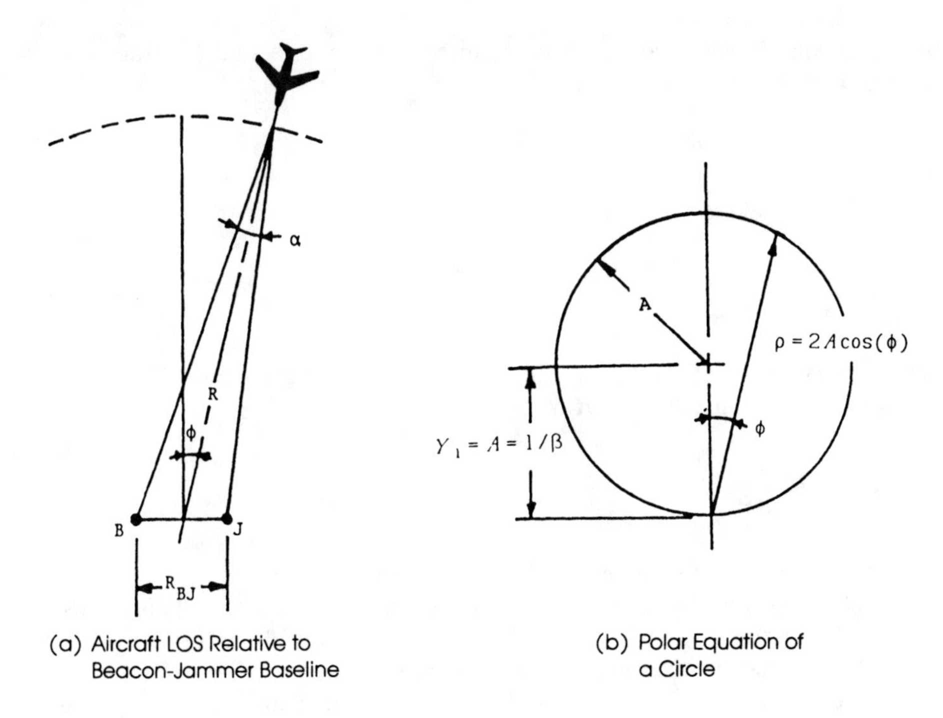

Figure 1.5 Derivation of boundary of main-lobe region.

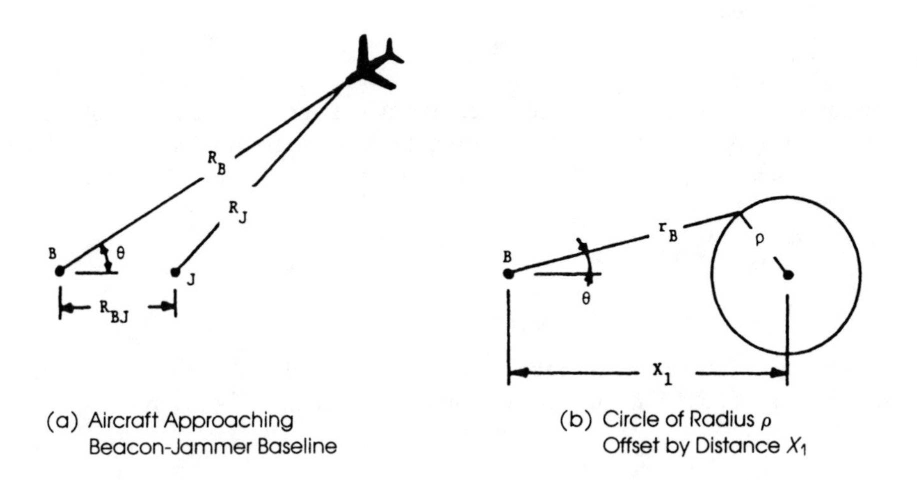

Figure 1.6 Derivation of sidelobe jamming contours.

By comparing the last two equations term by term, we see that (1.8) describes a circle of radius:

$$\rho = k|x_1| \tag{1.10}$$

the center of which is offset by

$$x_1 = \frac{1}{1 - k^2} \tag{1.11}$$

For given values of ERP_B, ERP_J, and SLR, a contour of constant[3] J/S is, as (1.6) states, a locus of constant $k = R_J/R_B$:

$$J/S = \left(\frac{\mathrm{ERP}_J}{\mathrm{ERP}_B}\right)\left(\frac{1}{\mathrm{SLR}}\right)\left(\frac{1}{k^2}\right) \tag{1.12}$$

Thus, a family of k-circles is a family of constant J/S contours with radii and center locations given by (1.10) and (1.11), respectively. Such a family is plotted in Figure 1.7, together with a few β-circles for several beamwidths. Inside the appropriate β-circle, the k-circles are loci of constant sidelobe-jamming J/S as given by (1.12). Outside the β-circle, the k-circles are loci of constant main-lobe-jamming J/S as given by (1.5), rewritten as

$$J/S = \left(\frac{\mathrm{ERP}_J}{\mathrm{ERP}_B}\right)\left(\frac{1}{k^2}\right) \tag{1.13}$$

The family of circles consists of two groups: those for which $k > 1.0$, and those for which $k < 1.0$. They are separated by the vertical line (circle of infinite radius), corresponding to $k = 1.0$, bisecting the beacon-jammer baseline. Each circle of the first group ($k > 1.0$) has a counterpart in the other ($k < 1.0$) corresponding to the k value of

$$k' = \frac{1}{k}$$

The two circles have identical radii. Note that the circles of the first group have negative values of x_1. This means that the center location of a given circle of this

[3]The constant J/S arises from the assumption of constant sidelobe gain for the radar antenna and constant ERP_B and ERP_J.

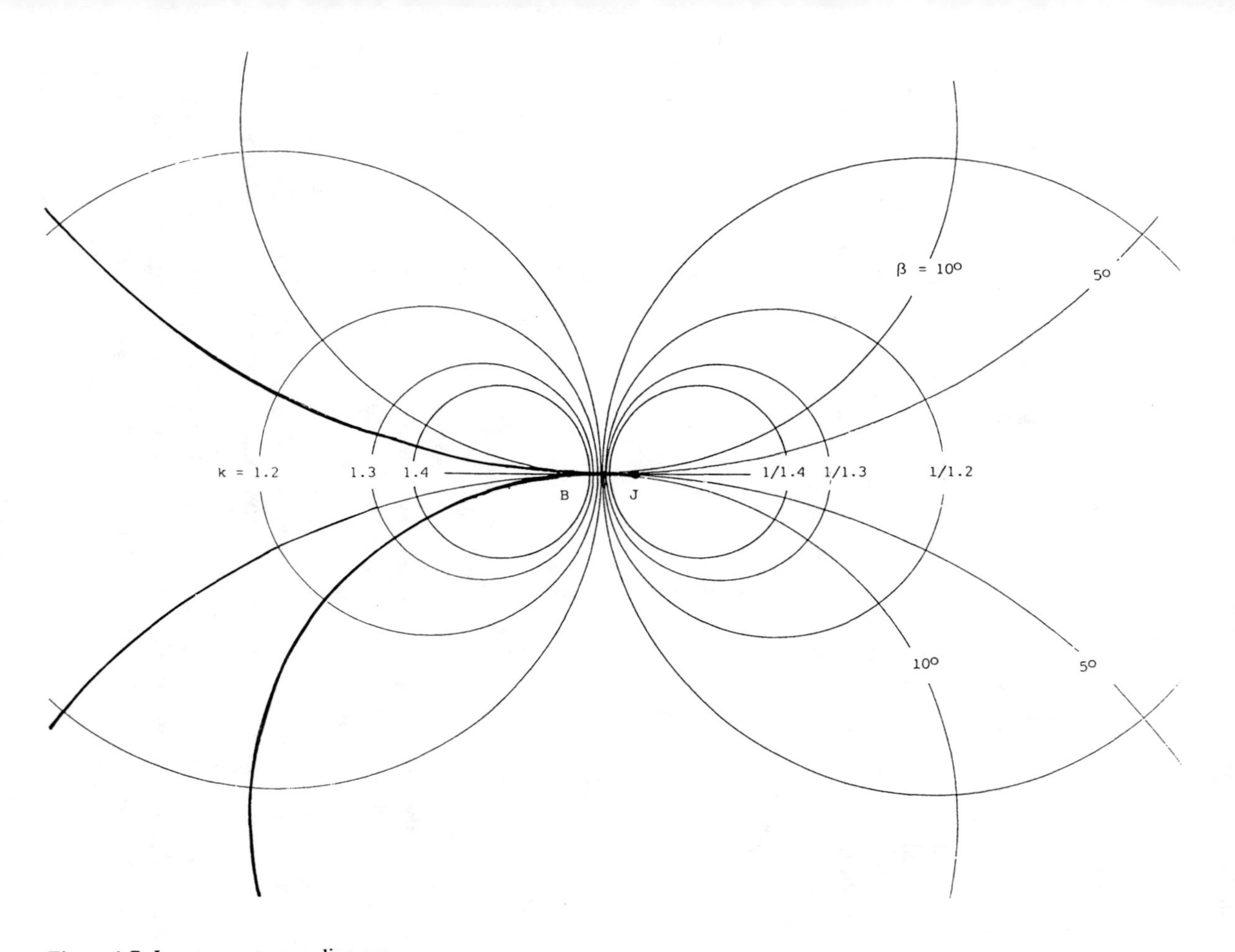

Figure 1.7 Jammer coverage diagram.

group lies to the left of the beacon (point B). Its counterpart has a center location given by

$$x_1' = 1 - x_1$$

which means that its center lies a distance x_1 to the right of the jammer location (point J).

Figure 1.7 is useful in the uplink jamming case because it describes the J/S intensity pattern in the space about the fixed baseline, R_{BJ}, at the ends of which the competing emitters are located. Such a diagram could be drawn for downlink jamming (jammer attacking the beacon receiver), but it would not be very useful. In the downlink case, the jammer is competing against the airborne radar transmitter, so the baseline now becomes the LOS path between jammer and radar. This baseline changes in length and orientation as the aircraft approaches. A more useful description for the downlink jamming case is a straightforward J/S equation involving the propagation path lengths, R_{BJ} and R_B, of the two competing emitters. This equation is

$$J/S = \left(\frac{\text{ERP}_J}{\text{ERP}_R}\right)\left(\frac{R_B^2}{R_{BJ}^2}\right)\left(\frac{G_{BJ}}{G_{BR}}\right) \tag{1.14}$$

where G_{BJ} and G_{BR} are, respectively, the beacon antenna gains in the directions of the jammer and the airborne radar. Because the beacon antenna pattern is so broad, the ratio G_{BJ}/G_{BR} is unlikely to give the beacon much directional selectivity advantage against the jammer. Nevertheless, the ground basing of beacon and jammer may put the jammer to a disadvantage not accounted for in (1.14). One factor is the likelihood that intervening terrain or vegetation may interfere with the propagation of the jamming. This is less likely in the uplink jamming case. Another factor is the deliberate placement of an obstruction near the beacon on the FEBA (Figure 1.4) side, thereby altering the beacon antenna pattern.

The jammer might use either noise or pulsed jamming against the beacon. If the beacon is equipped with AGC, noise jamming can cause the AGC to lower the beacon receiver gain so far that the beacon fails to respond to radar interrogations. Pulse jamming produces false interrogations and, if frequent enough, may occupy the beacon's attention to such an extent that the beacon fails to respond to the radar.[4] The false interrogations produce beacon emissions that can appear to a scanning radar to be coming from beacons at azimuths other than the true beacon azimuth (false replies received in radar antenna sidelobes). If the false replies are

[4]Generally, there is a refractory period after a beacon reply, during which time the beacon transmitter cannot be retriggered.

randomly timed (timing unrelated to radar interrogations), the radar can reject the false replies, because only valid replies, generally many per beam dwell duration, will integrate up to a level exceeding the radar's detection threshold.

Radar Beacon Jamming (Airborne Beacon)

The use of airborne beacons for air traffic control was mentioned earlier. A beacon on an incoming aircraft provides aircraft identity information and a clutter-free display and measurement of aircraft position for *ground-controlled approach* (GCA) purposes. It is possible that in a wartime environment jamming would be directed at either the uplink or the downlink. In the GCA case, the jammer is likely to be at a considerable range disadvantage, for the beacon-radar range would be short. Jamming would more likely be effective against air-route surveillance operations. In any event, the uplink and downlink jamming equations, (1.5), (1.6), and (1.14), are applicable.

Surface-to-air missiles (SAMs) may also carry beacons. Figure 1.8 depicts an attack aircraft A entering a region protected by SAM defenses. At least in the early phases of the engagement, the missile depends on command guidance from the launch area. The radar might skin-track the missile, but a more dependable track is obtained if the missile M carries a beacon B with an aft-looking antenna. The beacon replies provide missile position information to the radar, and, with appropriate coding, they can carry missile status information, such as missile attitude, altitude, acceleration, and control-surface orientation. The air crew, alerted by their radar warning receiver, may initiate jamming against the SAM uplink or downlink or against the target-tracking function. Figure 1.8 implies that a single radar performs both missile and target tracking,[5] but separate radars may be used for these two functions. Beacon interrogation and target-tracking frequencies may or may not be the same. If they are the same, jamming aimed at the missile-tracking link may also be effective against the target-tracking function.[6] Presumably, the chosen jamming tactic would have been determined before the mission by the choice of jammers allocated and by the preprogramming of any programmable jammers. Such choices would be made on the basis of estimated jamming vulnerabilities of the various elements of the enemy's SAM system and on estimates of attainable J/S values computed from radar and beacon jamming equations of the kind under discussion. Note that the aft-looking beacon antenna makes it difficult to jam the uplink of the SAM's missile guidance system.

[5]An electronically steered array antenna can track multiple targets.

[6]The broad antenna pattern of the airborne self-protection jammer would encompass both the launch site and the missile in flight, at least in the early phase of the engagement.

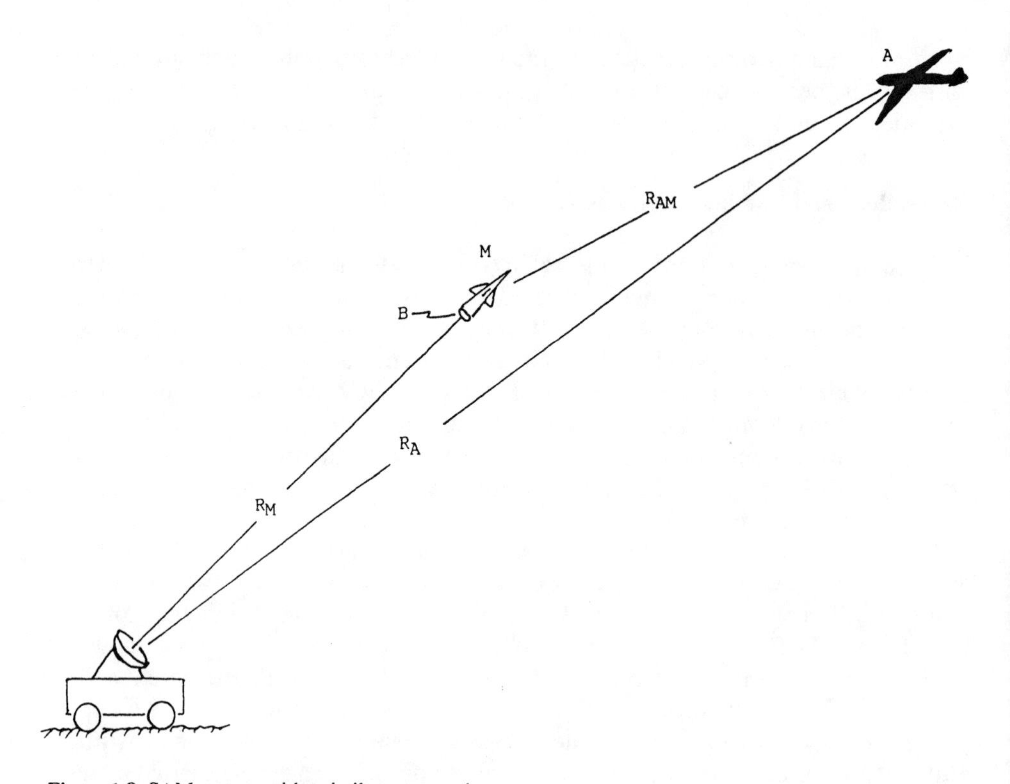

Figure 1.8 SAM system with missile transponder.

1.4 ACCOUNTING FOR ATMOSPHERIC ABSORPTION

In the analysis so far, the only path loss accounted for is the spherical wavefront spreading loss, which is proportional to $1/4\pi R^2$ for a one-way path and to $1/(4\pi R^2)^2$ for a round-trip. In clear weather and at frequencies below X band, this approximation is likely to be adequate. In fog, in precipitation, and at shorter wavelengths, atmospheric absorption [2, Chapter 2.4] must be taken into account.

Atmospheric absorption causes a propagating wave's field strength (and its power density) to decay exponentially with distance R. The power density decay factor is of the form:

$$L_a = e^{\alpha R}$$

In this definition, $L_a \geq 1.0$; it is the ratio of power density at the beginning to power density at the end of a path of length R, exclusive of the effects of the spreading loss. Published path-loss data often give round-trip loss for radar calculations. If so, one must take care to convert to one-way values when dealing with one-way propagation. Thus, if the loss is given in dB per unit of round-trip path distance, then, for a one-way path, we must divide the dB value by 2 (the equivalent of taking the square root of a power ratio). If L_a for any situation is appreciably greater than 1, we should take it into account. In all the equations of the preceding sections, jammer-to-signal ratios are expressed as numerical power ratios, not in dB. Thus, if we extract a value $(L_a)_{dB}$ from published dB loss data, we must convert it to a power ratio L_a for use in those equations:

$$L_a = 10^{0.1(L_a)_{dB}}$$

Then, to account for the atmospheric loss, we need only divide the loss-free received power (or power density) by L_a for one-way propagation and by L_a^2 for a round-trip.

For the escort-SPJ case, (1.1) corrected for atmospheric loss becomes

$$J/S = \left(\frac{\text{ERP}_J}{\text{ERP}_R}\right)\left(\frac{4\pi R^2}{\sigma_T}\right) L_a \tag{1.15}$$

Thus, the atmospheric absorption loss places the signal at a further disadvantage because it experiences the loss both going and coming, whereas the jamming experiences the loss only on the one-way transit. For the stand-off jamming case, (1.3) corrected for atmospheric loss becomes

$$J/S = \left(\frac{\text{ERP}_J}{\text{ERP}_R}\right)\left(\frac{R_T}{R_J}\right)^2\left(\frac{4\pi R_T^2}{\sigma_T}\right)\left(\frac{1}{\text{SLR}}\right)\left[\frac{(L_a)_T^2}{(L_a)_J}\right] \tag{1.16}$$

In (1.16), $(L_a)_T$ is the one-way atmospheric loss along path R_T, and $(L_a)_J$ is the one-way loss along R_J. For the radar beacon jamming scenario, the main-lobe jamming equation, (1.5), corrected for atmospheric loss becomes

$$J/S = \left(\frac{\text{ERP}_J}{\text{ERP}_B}\right)\left(\frac{R_B^2}{R_J^2}\right)\left[\frac{(L_a)_B}{(L_a)_J}\right] \tag{1.17}$$

where $(L_a)_B$ and $(L_a)_J$ are, respectively, the one-way atmospheric losses along paths R_B and R_J of Figure 1.4. The sidelobe jamming equation (1.6) becomes

$$J/S = \left(\frac{\text{ERP}_J}{\text{ERP}_B}\right)\left(\frac{R_B^2}{R_J^2}\right)\left(\frac{1}{\text{SLR}}\right)\left[\frac{(L_a)_B}{(L_a)_J}\right] \tag{1.18}$$

1.5 ACCOUNTING FOR MULTIPATH PROPAGATION EFFECTS

The models upon which this analysis rests presume a single point-to-point propagation path. So far, the possibility of multipath effects resulting from reflection or scattering of electromagnetic energy off the earth's surface has been ignored. To take these effects into account is not a simple matter, for the strength of the reflected component depends on wavelength, incidence angle, polarization, and the conductivity, dielectric constant, and roughness of the reflecting surface. Nevertheless, the phenomenon is well documented [5, Chapter 6]. Given reasonable estimates of the pertinent parameters, one can estimate the depths of the nulls in the vertical lobing pattern that result from multipath [2, Chapter 2]. At a null location, the reflected signal partially cancels the signal received via the direct path. The *null depth* is defined as the ratio of the power received in the null to the power that would be received in the absence of multipath reflections. The null depth is a measure of the jamming power loss that results from the destructive interference caused by multipath. A mission planner should be aware of this effect and should assume that there will be positions where the jamming power reaching the victim radar will be reduced by mutipath effects by a factor commensurate with the predicted null depths.

1.6 ACCOUNTING FOR WASTED JAMMING POWER (POWER FALLING OUTSIDE THE VICTIM RADAR'S PASSBAND OR RANGE GATE)

A radar is capable of operating over a frequency band within which its center frequency (the carrier frequency) can be tuned. The tunable band may span hundreds of megahertz or even a few gigahertz. Some radars have pulse-to-pulse frequency agility, meaning that the radar is capable of retuning in the dead time between pulses. The agility band may or may not span the entire operating band. The *intermediate frequency* (IF) section of the radar's superheterodyne receiver has a fixed center frequency and an acceptance passband comparable to the width of the spectrum of the radar return. Radars with selectable pulsewidths commonly have matching selectable IF passband widths.

The question of jammer spectrum width was not raised until this point in the analysis. In developing the J/S equations, we tacitly assumed that the jamming power would get through the radar receiver to accomplish its intended purpose, and that the incident J/S therefore had a meaningful relation to J/S further down the receiver chain. In deception jamming, this assumption is reasonable, for normally the deception waveform is a false-target waveform, designed to resemble a real-target return. However, this assumption is not reasonable in noiselike jamming, especially if the jammer operates in a barrage mode with the intent of jamming a frequency agile radar or multiple radars operating at different frequencies anywhere within a wide tunable bandwidth. In this case, the jammer emits a broad energy spectrum, getting part of the energy into one radar receiver and part into another. Clearly, the total power represented by J is not effective on any radar. Even if a single radar is involved, the jammer spectrum is normally widened beyond the known IF passband width of the victim radar to allow for uncertainties in jammer and radar tuning. Ideal band-limited noise has a white spectrum; that is, the noise power spectral density is uniform across the spectrum. If a jammer emits white noise of bandwidth B_J completely overlapping a radar acceptance band of width B_R ($B_R < B_J$), only the fraction B_R/B_J of the incident jamming power gets through the radar receiver. The remainder is ineffective, except in the unlikely event that the jammer's intent is to achieve overload or burnout of the radar receiver's front-end components (e.g., mixer). Thus, the J/S of interest includes only the jamming power within the radar receiver passband. When we apply any of the foregoing J/S equations to a noise-jamming situation, we must multiply J/S by B_R/B_J if the ERP_J factor has been defined to include the total noise-barrage jamming ERP. The alternative is to include in the definition of ERP_J only the portion of the noise power lying within the passband of the victim radar receiver.

At the beginning of this section, we suggested that waveforms for deception generally mimic the waveform of a real-target return. When this is so, the spectrum width B_j of the jamming waveform is equal to the target return signal spectrum width B_s. The radar receiver passband width, B_R, is matched to B_s, so $B_R/B_s = B_R/B_j = 1$, and we can conclude that there is no wasted jamming power. This conclusion is correct only if the receiver channel is ungated, accepting radar returns from all ranges, as is the case for surveillance radar. In a tracking radar, the receiver channel is range-gated and accepts signals only during the gate interval, which is aligned in time with the expected target return. If the deception-jamming pulse of duration τ_j lies within the gate interval, the pulse is accepted, but if its timing is incorrect, part of the jamming pulse may be deleted by the gate; its effective duration is reduced, so the pulse energy accepted by the gate is reduced by the factor T_{OL}/τ_j, where T_{OL} is the duration of overlap of the jamming pulse with the range gate. As T_{OL} becomes appreciably less than τ_j, the shortened pulse may be viewed, insofar as the receiver response is concerned, as a quasi-impulse of strength proportional to T_{OL}/τ_j. The effective jamming power is therefore reduced by that factor.

Sometimes, jamming pulses aimed at creating confusion for a surveillance radar may not be matched in duration to the target return. A *continuous-wave* (CW) jamming signal of constant amplitude may have its carrier frequency swept, in a deterministic or random pattern, across some radar band, with the intent of injecting a burst of jamming power into any radar receiver with passband of width B_R crossed by the sweep. The effect on the radar is the same as a pulse of duration:

$$\tau_j = \frac{B_R}{\dot{f}}$$

where $\dot{f}$ is the sweep rate (e.g., $\dot{f}$ could be in MHz/s if B_R is in MHz). One might expect that the full effect of the jammer would be felt if τ_j corresponds to the duration, τ_s, of a target return pulse. This proves to be the case even though the jamming pulse contains a linear frequency modulation on its carrier. Setting $\tau_j \geq \tau_s$ yields

$$\dot{f} \leq \frac{B_R}{\tau_s}$$

For simple pulse waveforms, $B_R \approx 1/\tau_s$, so the preceding inequality becomes

$$\frac{\dot{f}}{B_R^2} \leq 1$$

As long as the jammer sweep rate satisfies this relationship, a full-amplitude jamming pulse is produced at the output of the radar's IF amplifier. As has been shown [6, Chapter 8], the pulse decreases according to the following relation as the sweep speed increases:

$$\text{relative amplitude} = \left[1 + 0.195\left(\frac{\dot{f}}{B_R^2}\right)^2\right]^{-1/4}$$

This relation is based on the assumption that the filter has a Gaussian-shaped passband.

1.7 ACCOUNTING FOR SIGNAL-PROCESSING GAIN

The J/S equations of Section 1.3 are expressed in terms of ERP ratios. These ratios pertain to the signal and jammer powers leaving the radar receiving antenna and entering the receiver. The equations say nothing about the possible effects on J/S of filters and other elements of the receiver chain, nor of the effects of signal processing or digital processing that may occur beyond the receiver proper. Figure 1.9 is a generic radar receiver block diagram indicating signal flow through the receiver to a display, to range and angle trackers, and to a digital data processor. Depending on the radar's purpose, some elements may be missing. A surveillance radar needs no range, angle, or doppler trackers. A noncoherent radar can have no coherent detector. If the radar employs a simple pulse waveform, the IF amplifier of proper bandwidth serves as the (approximately) matched filter. The target detection func-

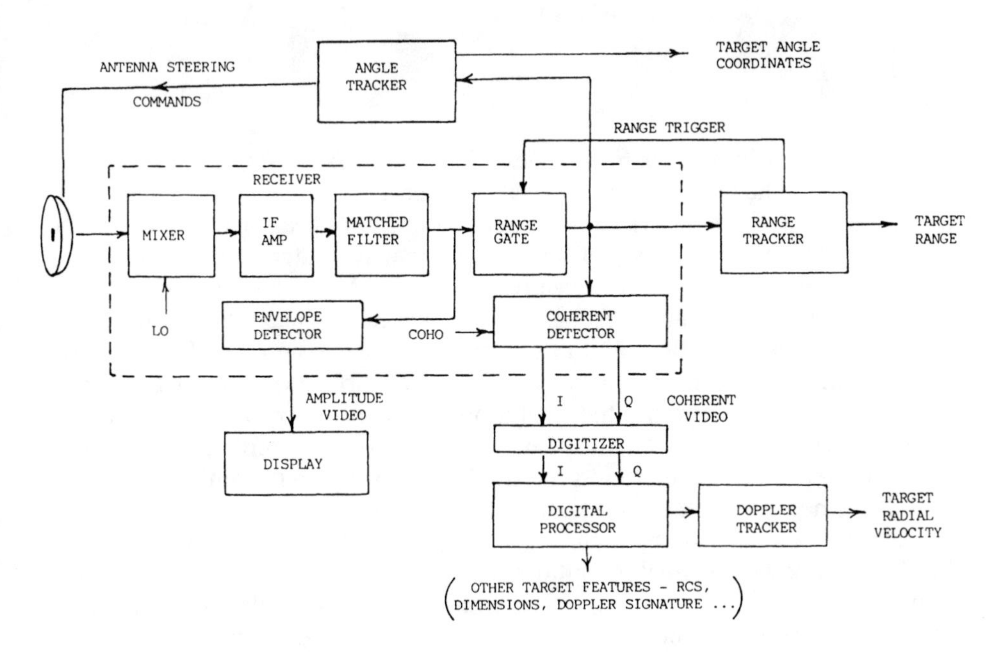

Figure 1.9 Generic pulse radar receiver block diagram.

tion is not explicitly shown. It may be performed manually on a display, by an analog threshold device operating on the envelope-detected video, or digitally on digitized range-gated video. In the latter case, a bank of contiguous range gates is required to process the returns from the range swath that defines the surveillance area.

We shall see that the block in Figure 1.9 labeled "matched filter" can change the jammer-to-signal ratio as the signal and jamming pass through it. In a benign environment (jamming absent), the target return competes against ever-present system noise. The signal-to-noise ratio (S/N) is of great concern when weak targets must be detected or tracked in system noise [7, Chapter 4]. The *matched filter* is the filter that maximizes the output S/N.

In the preceding section, we discussed the simple pulse of duration τ_s, the spectrum width of which is approximately,

$$B_s = \frac{1}{\tau_s} \tag{1.19}$$

and we remarked that the receiver passband should be

$$B_R = B_s = \frac{1}{\tau_s}$$

If the passband width is appreciably less than B_s, a significant amount of signal power is lost. If the passband is appreciably wider, excessive noise power is admitted. Equation (1.19), together with a requirement for a linear phase characteristic, constitutes the approximate specifications for a matched filter for the simple pulse.

Normally, we think of the range-resolution capability (potential for revealing the separateness of two targets that are closely spaced in range) of a signal as being determined by pulsewidth τ, and this is the case for the simple pulse waveform. Nevertheless, it has been demonstrated [8] that signal bandwidth is the fundamental parameter that determines the resolution potential of a waveform. Because of (1.19), one can make the equally valid claim that, for the simple pulse, the pulse duration determines the resolution potential. To constrain a transmitted waveform by (1.19) is not essential. By modulating the frequency or the phase of the carrier within the pulse, one can make the pulse spectrum width much greater than $1/\tau_s$, thereby making the potential resolution much finer than the resolution potential of a simple pulse of equal duration. Pulses with intrapulse modulation of the carrier are called *coded pulses* (the code is inherent in the modulation pattern).

Often a radar transmitter is operated in a saturated (maximum output) mode, in which case the pulse energy can be increased only by extending the duration of the pulse. With range resolution determined by the intrapulse modulation, the radar designer is free to employ long pulses. Many modern radars transmit long coded pulses with BT products (product of pulse duration T and pulse bandwidth B) much greater than 1. Upon reception, the pulse is "compressed" to a duration:

$$\tau_c = \frac{1}{B}$$

This compression is brought about by a matched filter. The voltage transfer function of the matched filter is the complex conjugate of the amplitude spectrum of the coded pulse waveform to which it is matched. The compression ratio is the ratio of the uncompressed-pulse duration to the compressed-pulse duration:

$$\text{compression ratio} = \frac{T}{\tau_c} = \frac{T}{1/B} = TB$$

Because all the pulse energy is compressed into the narrowest possible packet, the pulse power level of the compressed pulse is the maximum possible for a given pulse energy E. Receiver noise has no systematic structure. The matched-filter pulse compressor has no effect on the noise statistics, so the output rms noise power remains equal to the input noise power. Therefore, S/N at the matched filter output is maximized.

If we assume the matched filter is lossless, the pulse energy at the input equals that at the output. If P_u and P_c are, respectively, the pulse power levels of the uncompressed and compressed pulses, we can write

$$E = P_u T = P_c \tau_c$$

In terms of power, we can say that the pulse compression process results in a power gain, g_{p1}, equal to the compression ratio:

$$g_{p1} = \frac{P_c}{P_u} = \frac{T}{\tau_c} = BT$$

The subscript 1 associated with g_p indicates that this is a single-pulse processing gain. Later, we consider the processing of pulse trains containing many pulses. The signal-to-noise ratio at the matched-filter output is greater than the input S/N by a factor equal to the processing gain:

$$(S/N)_{\mathrm{out}} = g_{p1}(S/N)_{\mathrm{in}}$$

Therefore, if a jammer transmits a noiselike waveform against a pulse-compression radar, thereby achieving a jammer-to-signal power ratio, $(J/S)_{\mathrm{in}}$, at the radar receiver input, J/S at the output of the pulse compressor, $(J/S)_{\mathrm{out}}$, clearly will be lower by a factor equal to the reciprocal of the processing gain (the signal becomes larger relative to the jamming):

$$(J/S)_{\mathrm{out}} = \frac{1}{g_{p1}} (J/S)_{\mathrm{in}}$$

The same will be true of any jamming waveform that lacks the intrapulse structure of the signal. A jammer that amplifies and plays back a copy of the incident radar waveform is called a *repeater jammer.* Clearly, unless the repeater introduces appreciable distortion, the repeated waveform will enjoy the same processing gain, g_{p1}, as a target return, with the result that J/S is unchanged in passing through the radar's pulse-compression filter. For simple pulses ($BT = 1$), there is no pulse-compression gain; perhaps, more properly, we should say that the gain is 1. Therefore, for any jamming waveform of bandwidth equal to the simple pulse bandwidth, J/S is unchanged in passing through the radar receiver.[7]

[7]This statement and the preceding statements relating input and output J/S assume use of a linear receiver.

In general, more than one target return pulse is received from a target during the beam dwell, T_β. For a mechanically scanning surveillance radar with beamwidth β and scan rate Ω, the dwell is

$$T_\beta = \frac{\beta}{\Omega}$$

A radar with electronic beam steering, unhampered by mechanical inertia, can adjust its T_β to collect the desired number, n_β, of pulses per dwell. With a *pulse repetition frequency* (PRF) equal to f_R, the number of pulses per dwell is

$$n_\beta = f_R T_\beta$$

For detection of target returns embedded in system noise, to base the detection decision on the sum of the n_β pulses is advantageous. Formation of the sum is called *integration.* It amounts to overlaying (stacking) successive received range sweeps and adding them point by point. In a coherent radar, the summation is performed before detection (integration at IF), which is known as *coherent integration.*[8] Figure 1.10 aids in explaining the benefits derived from multipulse coherent integration. Figure 1.10(a) shows the n_β target return voltages, each of magnitude v_{s1}, adding up to a total integrated signal voltage:

$$v_s = n_\beta v_{s1}$$

The phasors add in phase because the radar and the target are stationary. In Figure 1.10(b) the target return phasors exhibit a progressive phase shift of increment $\Delta\phi$ from a one pulse repetition interval to the next, because of the target's radial motion (the progressive phase shift corresponds to a doppler shift).

Analog and digital methods have been devised for integrating phasor sequences with linear phase progressions. The most common method now in use applies the *fast Fourier transform* (FFT) process to the range-gated, digitized sequence of coherent video pulses. The FFT processor has n output channels, each with a different internal phase progression that is applied to the sequence of input pulses. The channel with the phase progression that compensates the incoming pulse phase progression yields an output voltage equal to $n_\beta v_{s1}$, just as large as though the target had been stationary. Incidentally, the channel number in which this peak signal occurs reveals the doppler shift (and radial velocity) of the target. Figure 1.10(c) shows the phasor sum of the system noise voltage samples v_{n1}, v_{n2}, v_{n3}, . . . that accompany the target returns. The phasor addition of these noise sam-

[8]Postdetection integration can be coherent, provided that a coherent detection process is employed.

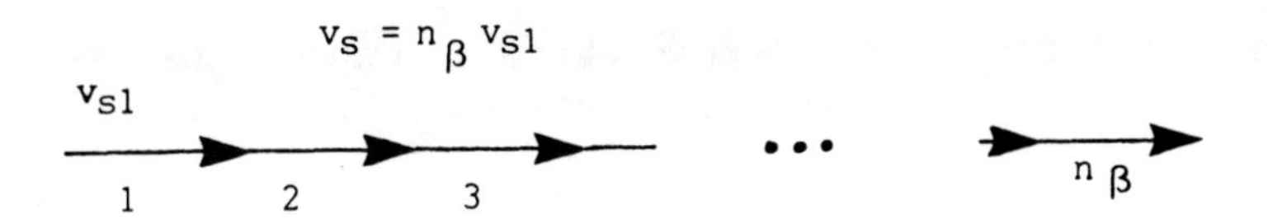

(a) n_β Returns From Stationary Target

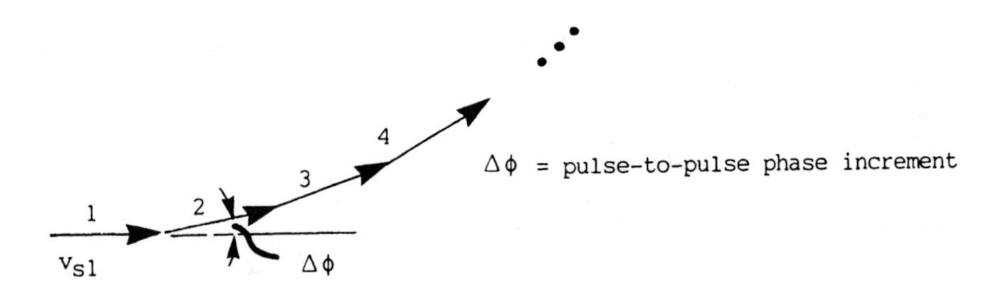

(b) Returns From Moving Target

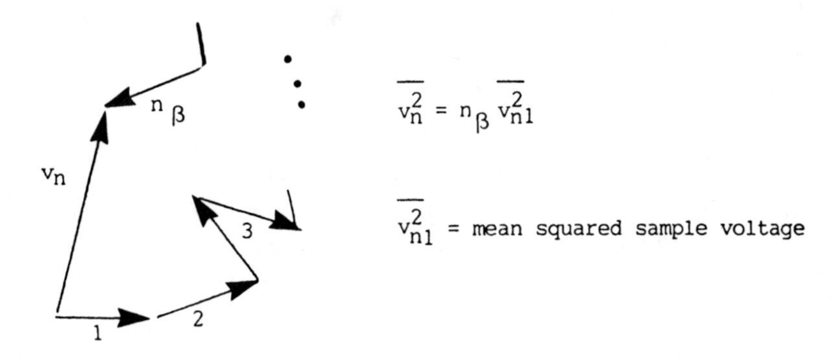

(c) Summation of n_β Noise Samples

Figure 1.10 Multiple pulse coherent integration.

ples is equivalent to the two-dimensional random walk [9, Chapter 3] outcome. The conclusion is that the noise power of the summation is

$$P_n = n_\beta P_{n1}$$

where P_{n1} is the rms noise power associated with a single noise sample (it is the noise power against which a single target return competes for detection).

The signal power gain (the square of the signal voltage gain) is

$$\left(\frac{v_s}{v_{s1}}\right)^2 = n_\beta^2$$

Therefore, the processing gain attributable to the n-pulse coherent integration is (recall that S and N are powers)

$$g_{pn} = \frac{(S/N)_{\text{out}}}{(S/N)_{\text{in}}} = \left(\frac{S_{\text{out}}}{S_{\text{in}}}\right)\left(\frac{N_{\text{in}}}{N_{\text{out}}}\right) = (n^2)\left(\frac{1}{n}\right) = n$$

If a jammer transmits a noiselike waveform against a coherent radar (n-pulse integration), achieving a ratio $(J/S)_{\text{in}}$ at the radar receiver input, then $(J/S)_{\text{out}}$ at the output of the coherent integrator will be lower by a factor equal to the reciprocal of the n-pulse processing gain:

$$(J/S)_{\text{out}} = \frac{1}{g_{pn}}(J/S)_{\text{in}} = \frac{1}{n}(J/S)_{\text{in}}$$

Note that the n-pulse processing gain is independent of the single-pulse processing gain discussed earlier. Thus, a pulse-compression radar with n-pulse coherent integration achieves a total processing gain of

$$g_p = g_{p1}g_{pn}$$

The overall effect against a noise jammer, accounting for both the intrapulse processing gain and the n-pulse integration gain, is

$$(J/S)_{\text{out}} = \frac{1}{g_{p1}g_{pn}}(J/S)_{\text{in}}$$

Not only the noise jammer suffers the $1/n$ disadvantage against the coherent target return. A pulsed jammer with carrier phase unrelated to the phase of the incident radar pulse (a transponder) transmits pulses with random phase, and therefore suffers this same disadvantage.

Many radars are noncoherent, and multipulse integration is often noncoherent: that is, the pulses that are integrated are envelope-detected video. Noncoherent integration is less effective (it yields a lower value of integration gain) than coherent integration. Instead of achieving a gain equal to the first power of n_β, the gain of the noncoherent integrator is sometimes approximated [10, Chapter 1] by

$$g_{pn} = n^{0.8}$$

Note that integration achieved on the phosphor of a *cathode ray tube* (CRT) display is a familiar form of noncoherent integration.

1.8 ACCOUNTING FOR POLARIZATION LOSS

Chapter 4 (Section 4.6.2) discusses a special situation in which jamming energy is deliberately transmitted at a polarization orthogonal to the design polarization of the receiving antenna of the victim radar. This is uncommon; the normal preference is to match the victim radar's receiver polarization in order to maximize the jamming power coupled into the receiver. There are two possible reasons why this polarization match may not be feasible: (1) the radar's receiving polarization may not be known; and (2) the jammer may not have the capability to adjust its transmitting polarization. It is likely that the radar's receiving polarization is the same as its transmitting polarization, which can be observed as the incident radar signal polarization detected at the jammer. However, because most radar targets backscatter a significant amount of energy at polarizations other than the incident polarization, the radar is able to receive on a polarization different from its transmitting polarization. In any event, the ECM mission planner should be aware of the possibility of a polarization mismatch, and account for the resultant loss in estimating J/S levels.

For further reference on the ECM and antijamming scenario, the reader should consult Van Brunt [11] and Maksimov [12].

REFERENCES

1. Brookner, E., ed., *Aspects of Modern Radar,* Artech House, Norwood, MA, 1988.
2. Skolnik, M.I., ed., *Radar Handbook,* 1st Ed., McGraw-Hill, New York, 1970.
3. Roberts, Arthur, *Radar Beacons,* M.I.T. Radiation Laboratory Series, Vol. 3, McGraw-Hill, New York, 1948.
4. Scanlan, M.J.B., ed., *Modern Radar Techniques,* Macmillan, New York, 1987.
5. Blake, L.V., *Radar Range-Performance Analysis,* Artech House, Norwood, MA, 1986.

6. Wiley, R.G., *Electronic Intelligence: The Interception of Radar Signals,* Artech House, Norwood, MA, 1985.
7. Goldman, S., *Frequency Analysis, Modulation and Noise,* Dover, New York, 1967.
8. Woodward, P.M., *Probability and Information Theory with Applications to Radar,* Artech House, Norwood, MA, 1980.
9. Lawson, J.L., and G.E. Uhlenbeck, *Threshold Signals,* M.I.T. Radiation Laboratory Series, Vol. 24, McGraw-Hill, New York, 1950.
10. Barton, D.K., *Radar System Analysis,* Artech House, Norwood, MA, 1976. *See also* Barton, D.K., *Modern Radar System Analysis,* Artech House, Norwood, MA, 1988.
11. Van Brunt, L.B., *Applied ECM,* EW Engineering, Dunn Loring, VA, 1978.
12. Maksimov, M.V., *et al., Radar Antijamming Techniques,* Artech House, Norwood, MA, 1979.

Chapter 2
NOISE JAMMING

2.1 NOISE-JAMMING SEARCH AND ACQUISITION RADARS

The detection range of search and acquisition radars has been studied and analyzed extensively [1–4]. These studies considered thermal (random) receiver noise and fluctuations in target cross section. Propagation factors, such as multipath reflections, also received extensive consideration. The deterioration of radar performance from noise jamming is the subject of interest here.

The amount of noise power needed to obliterate a target is increased because of the fluctuation of target cross section. The occasional peaks in the cross section make the target visible above the jamming. If the target aircraft is self-screening and the defended point has a *home-on-jam* (HOJ) capability, J/S should be large enough to deny the radar the capability to measure range. Even an approximate range (e.g., ± 3 nmi) can be adequate for the successful launch of an HOJ missile. If range measurements can be denied the radar, the ability to launch such a missile is reduced.

To illustrate the difficulty of completely denying range, consider a target aircraft flying inbound at 400 knots, a *track-while-scan* (TWS) radar with 10 scans/s, and a probability of detection of only 1%. In the 54 s required to traverse 6 nmi, the target will present discernible echos on about 5 of the 540 scans. Thus, a probability of detection much lower than 1% (such as 0.01%) is desirable so that range measurement may be essentially eliminated.

As illustrated in Figure 2.1, the radar signal from the target increases as the inverse fourth power of range. This makes the obliteration of range information difficult at short ranges. The margin by which J/S exceeds the necessary level decreases as a function of range, as shown in Figure 2.2. On this graph, the J/S level for obliteration of range information is selected as 10 dB and represents a probability of detection of less than 0.01% for a target that fluctuates from scan to scan. Increasing the jammer power can decrease the minimum range for obliteration, which can be seen from the development of the J/S equations in Chapter 1.

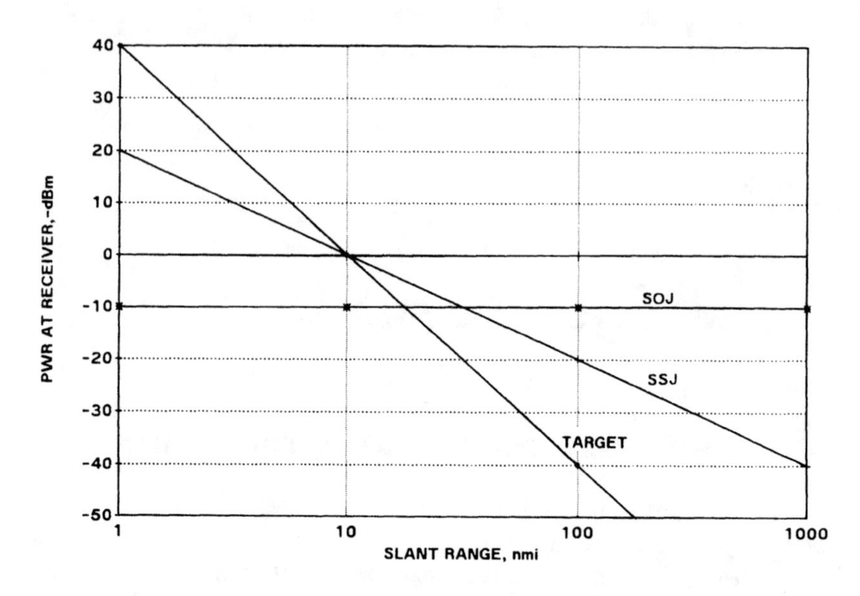

Figure 2.1 Jamming by a stand-off jammer (SOJ) or a self-screening jammer (SSJ).

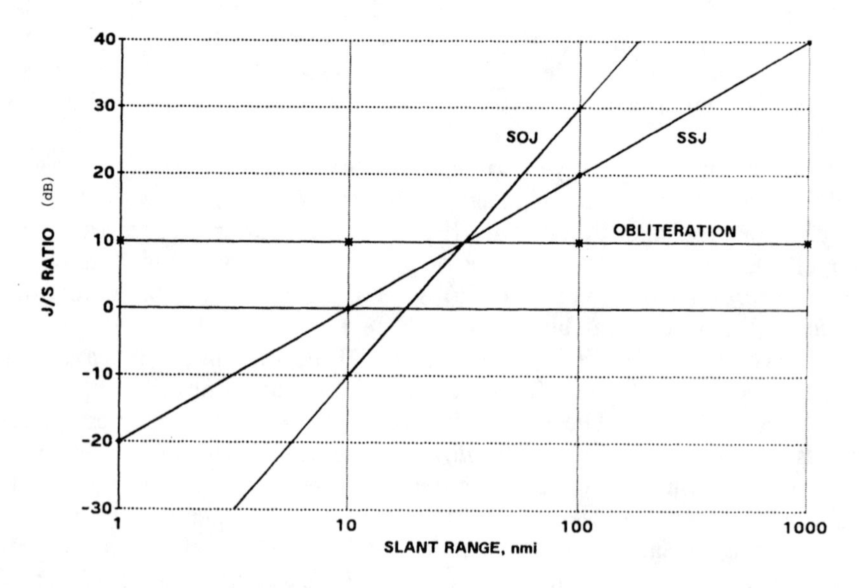

Figure 2.2 J/S as a function of range for SSJ and SOJ.

2.2 NOISE JAMMING A THRESHOLD DETECTOR

The effect of noise jamming on a radar that uses a threshold detector with a fixed threshold can be inferred from the standard curves for probability of detection and probability of false alarm. Usually, these probabilities are related to the integrated signal-to-noise ratio at the input to the threshold device. However, for jamming calculations, it is common to compute these probabilities in terms of the integrated jamming-to-signal ratio at the threshold input.

Consider an example. Suppose the radar threshold is set to provide $P_{FA} = 10^{-6}$. According to Figure 2.3, the threshold required is about 5.1 times the rms noise voltage. Now, suppose a noise jammer wishes to introduce additional noise such that the probability of false alarm increases to 0.5. This means that the threshold is now about 1.2 times the rms level of the jamming plus the noise. Since the threshold is fixed, this means that

$$\frac{J + N}{N} = \left(\frac{5.1}{1.2}\right)^2 \text{ or } 12.5 \text{ dB} \tag{2.1}$$

or

$$J \approx \left(\frac{5.1}{1.2}\right)^2 N - N \approx 17 \text{ N}$$

If the intent is to jam a search radar that is looking for a small target at its maximum range, the minimum jamming power required at the threshold input can be found rather easily. Suppose adequate radar performance occurs for an S/N of 10 dB. Then at the maximum range and for the smallest target the radar designer expects to have

$$S \approx 10 \text{ N}$$

To increase the false alarm rate to 0.5, we need $J = 17$ N or $J = 1.7S$. Clearly, if we make $J/S > 1$ at the threshold detector input, the radar's performance will be poor. Under these conditions, it is practical to neglect the receiver noise, which is typically at least 10 dB below both the signal to be detected and the jamming level. Thus, we can obtain the probability of detection and false alarm in the presence of jamming from Figure 2.3 by changing the sign and relabeling the S/N values as J/S values; for example, a $(J + N)/S$ value of 3 dB corresponds to the curve marked $S/N = -3$ dB. To determine precisely the false-alarm rate when noise jamming is present, we should take the receiver noise into account, especially if $J/S < 1$. Because the receiver noise and jamming noise are uncorrelated, their total power is simply $J + N$. This sum causes false threshold crossings, and one should

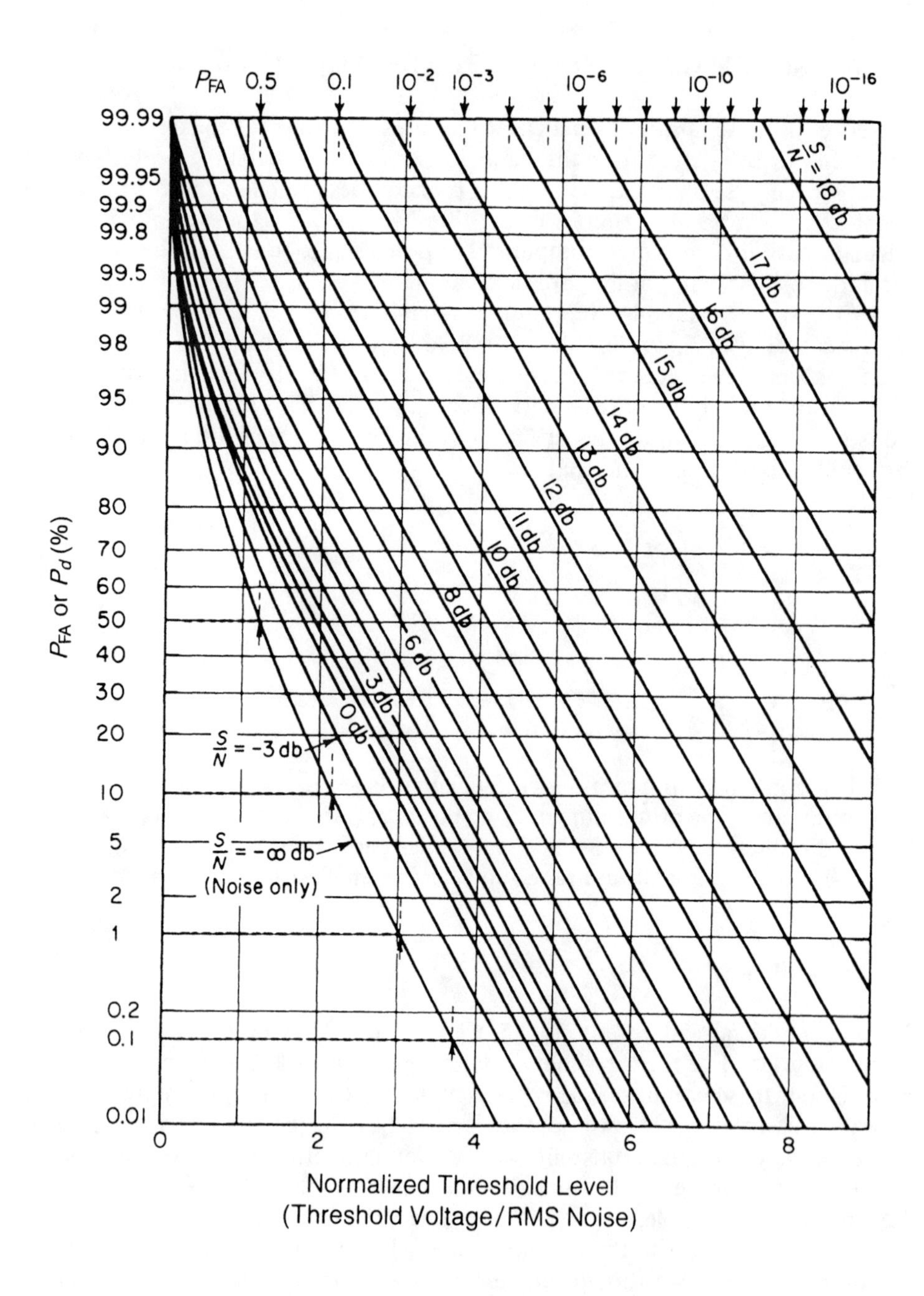

Figure 2.3 Probability of threshold crossing [1, p. 16]. *Note:* These curves are for a nonfluctuating target.

use it when determining the probabilities of detection and false alarm. The effect of this sum is illustrated in Table 2.1. For $J/S > 1$, the error due to ignoring the receiver noise is typically less than 0.5 dB, which is smaller than the uncertainties in target cross section, antenna patterns, range polarization losses, and other parameters that enter into J/S calculations in practice.

Table 2.1

J/S (dB)	S/N (dB)	$(J + N)/S$	$(J + N)/S$ (dB)
+10	10	10.1	10.04
+6		4.1	6.1
+3		2.1	3.2
0		1.1	0.4
−3		0.6	−2.2
−6		0.35	−4.6
−10		0.2	−7.0
+10	20	10.01	10.00
+6		3.99	6.01
+3		2.005	3.02
0		1.01	0.04
−3		.51	−2.9
−6		.26	−5.8
−10		.11	−9.6

Constant False-Alarm Rate (CFAR) Systems

The effect of noise jamming on radars that use *constant false-alarm rate* (CFAR) thresholding is to reduce the probability of detection while maintaining the same false-alarm probability. This happens because the threshold voltage increases in response to the jammer noise (we can also think of the gain ahead of the threshold decreasing in response to the noise). The effect on probability of detection is simply that of using the (reduced) $S/(J + N)$ with the standard detection curves. For example, suppose the CFAR threshold is set for $P_{FA} = 10^{-6}$. From Figure 2.3, this corresponds to about 5.1 times the rms noise voltage. If the jamming is sufficiently strong to provide a $(J + N)/S$ of 0 dB, the probability of detection is reduced to about 0.01%, which is practically negligible. Apparently, $J/S > 1$ at the threshold input is sufficient to jam the CFAR system. If the target is nonfluctuating for $J/S > 10$, there will be virtually no target visibility; hence, the term *obliteration jamming* is applied.

The effect of jamming on the CFAR system is to remove all threshold crossings due to targets; hence, there are no detections reported or displayed. Without CFAR, the effect is to create many additional threshold crossings and, thus, report or display many false targets. Of course, the human observer often knows that jamming is present. If a CFAR system is jammed, known clutter may disappear from the display. (If a non-CFAR system is jammed, the many false returns that appear at all ranges indicate the existence of strong jamming.) The J/S needed at the input to the CFAR-threshold circuitry to reduce the probability of detection to 0.01% is shown in Figure 2.4 as a function of *false-alarm number* (FAN). We obtain this graph by replotting selected points from Ref. 8 for a case 1 nonfluctuating target. (The probability of false alarm is 0.69/FAN.) As indicated, a J/S of +10 dB will reduce the probability of detection to 0.01% even if the FAN is reduced to 10^4 ($P_{FA} \approx 0.69 \times 10^{-4}$).

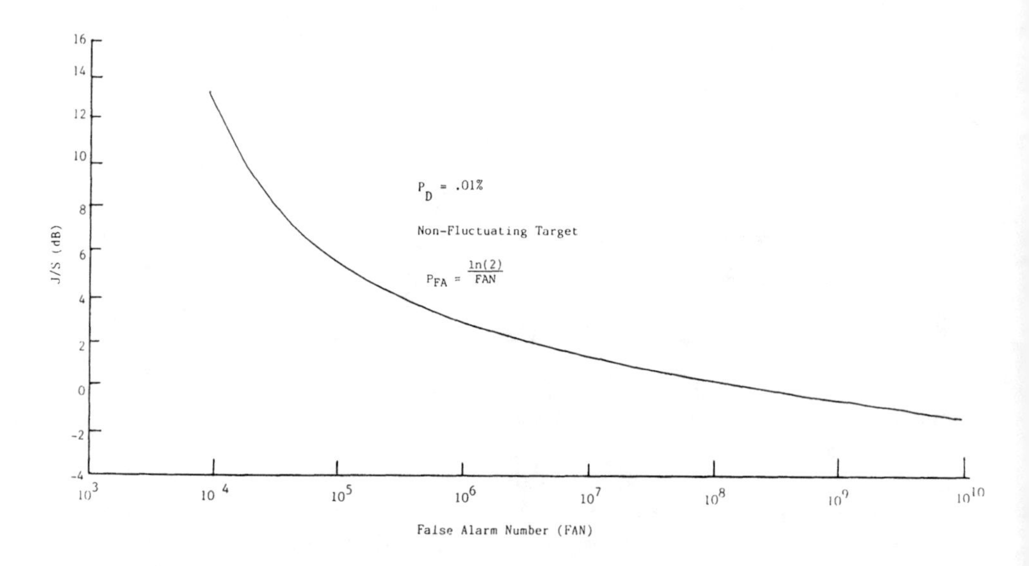

Figure 2.4 J/S required for $P_D = 0.01\%$.

Properly designed CFAR systems generally can detect the presence of jamming by monitoring the CFAR threshold. Then the direction of the jamming can be indicated even if the target echos are lost. The proper response of the jammer is to use "inverse gain" modulation of the jamming noise level. This means that the level of the jamming is inversely proportional to the signal being received from the radar. In this way, the jammer power entering the radar receiver is approximately constant regardless of the direction in which the radar beam is pointing (this assumes that the radar uses the same antenna gain on reception as on transmission). Of course, instantaneously monitoring the radar signal requires numerous "look-through" intervals when the jamming must be turned off. During these intervals, an ESM receiver that is integrated into the ECM system measures the amplitude of the signal from the radar. This may be referred to as a "smart noise" jamming technique.

2.3 FALSE-TARGET GENERATION USING REPEATED NOISELIKE SIGNALS

The use of white-noise-like jamming creates transitory false targets that quickly appear and disappear at random ranges and angles. Generating more realistic false targets requires the use of more intelligent noiselike signals. For example, if the same noiselike signal is repeated *exactly* during each radar *pulse repetition interval* (PRI), for approximately the same number of pulses integrated by the radar, then not only are the false targets more like real ones but the jamming is more effective because it uses the noncoherent integration gain of the radar to increase the jamming effect. Such jamming requires generation of pseudonoise sequences (e.g., by use of random number generators) or storage of an actual noise sequence, which is then repeated.

To compute the probability of generating a false-target response, we determine J/S before the noncoherent integration and then increase it by n^{γ}, where n is the number of repetitions of the noise integrated by the victim radar and γ is the noncoherent integration efficiency (typically, $\gamma = 0.5$ to 0.8). For a typical search radar, n should be comparable to the number of pulses per beamwidth transmitted by the radar. For a circular scan, this number is

$$\frac{\theta_{3\mathrm{dB}}}{360^\circ} \times \frac{T_S}{\mathrm{PRI}} \qquad (2.2)$$

where

$\theta_{3\mathrm{dB}}$ = azimuth 3-dB beamwidth of radar (degrees);
T_S = scan time (seconds) of radar antenna;
PRI = pulse repetition interval (seconds).

In addition, the precise PRI must be measured—it must be accurately determined to an accuracy of approximately

$$\Delta T < \frac{\text{pulse duration}}{N}$$

This determination is necessary to make the noise add properly over the sequence of N PRIs (assuming the noise bandwidth is approximately the reciprocal of the pulse duration). Then the noise mimics the echoes from random point targets at each radar beam position. Note that we do not want to create false-target responses in each range-angle-doppler cell, but only in about half of them. We must precisely determine PRI, pulse duration, and pulses per beamwidth before performing noise jamming of this type. Hence, this type of jammer requires an associated ESM receiver and processor to make the measurements. In addition, we need "look-through" intervals to check the victim radar's parameters; after all, if the radar were to change its PRI, the effectiveness of the jamming would be reduced—and the jammer would be unaware of its reduced effectiveness. Interception and analysis of radar signals (discussed in Refs. 5 and 6) involve precision measurements of radar signal parameters needed to conduct this kind of jamming successfully.

2.4 NONLINEAR CIRCUITS AND THEIR EFFECT ON J/S

A central problem in analyzing the vulnerability of a radar to noise jamming is to determine what J/S exists at the receiver output for a given value of J/S at the receiver input. This determination is difficult because the receiver may contain nonlinear devices, such as detectors and limiters.

A detector (e.g., linear or square-law) will usually be included at the output section of a radar receiver. The detector will give the sum of the squares of the in-phase and quadrature signal and noise components for square-law or the root of the sum for linear detection. For either detector, there is a small-signal suppression effect for large J/S. Davenport and Root [7, p. 306] give the effect of the detector on the output J/S as a function of the detector type (linear, square-law, etc.). Their general formulation of detector action includes provision for the exact probability distribution of the input jamming signal, although for this discussion they assume Gaussian white noise. For large values of J/S, the output J/S becomes proportional to the square of the input J/S, whereas for low values the proportionality is direct. In the transition region, Lawson and Uhlenbeck [8, p. 282] suggest the use of a detector factor, which is the right factor in the equation:

$$(J/S)_o = (J/S)_i \left(\frac{2S + J}{2S} \right)_d \tag{2.3}$$

where

o = output ratio subscript;
i = input ratio subscript;
d = detector factor.

Figure 2.5 is a plot of this detector factor. The curves show the small difference for linear and square-law detectors. The usual receiver noise, which is present along with any jamming signals, must be included. Modifying the detector factor to include the receiver thermal noise gives

$$\left(\frac{J+N}{S}\right)_o = \left(\frac{J+N}{S}\right)_i \left(\frac{2S+J+N}{2S}\right)_d$$

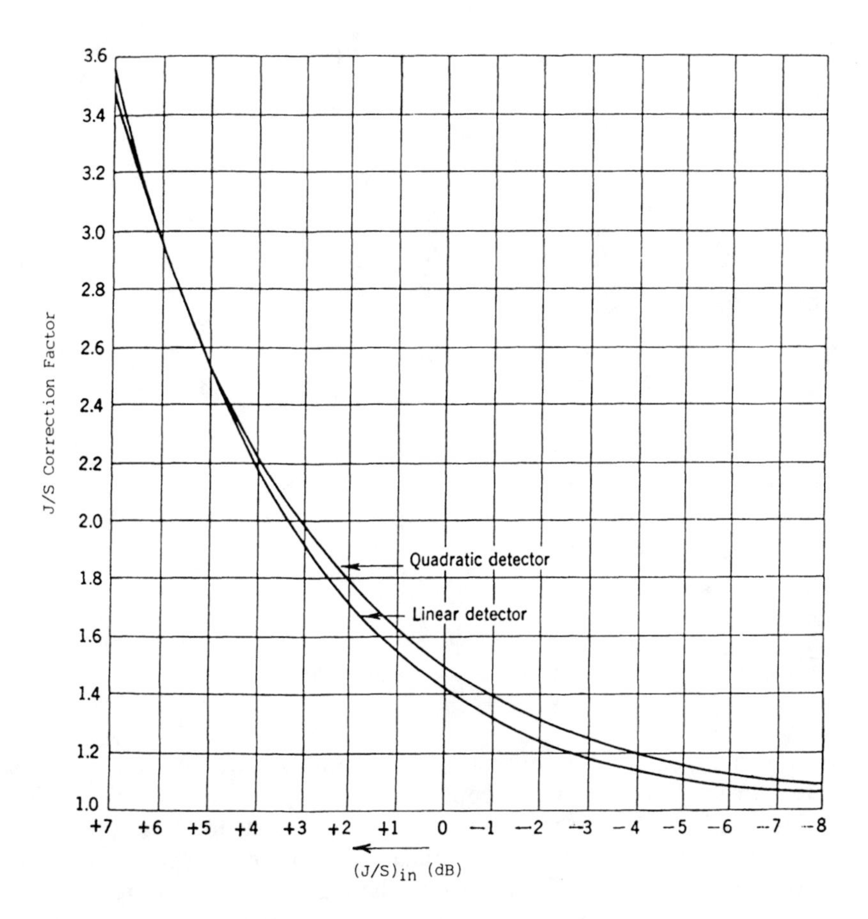

Figure 2.5 Detector factor as a function of (J/S) in [8, p. 283].

Curves for probability of threshold crossing and probability of detection as a function of false-alarm probability customarily include the effect of a detector [1, p. 16, 17], and may include provision for noncoherent integration [4]. Here, the effect of noncoherent integration is to be accounted for in the receiver's processing gain.

If the amplification in the receiver is nonlinear (such as inadvertent limiting, logarithmic, or hard limiting), $(J/S)_o$ can be altered. Consider the effect of ideal limiting on $(J/S)_o$. The ideal limiter has a symmetrical, rectangular transfer characteristic, as shown in Figure 2.6.

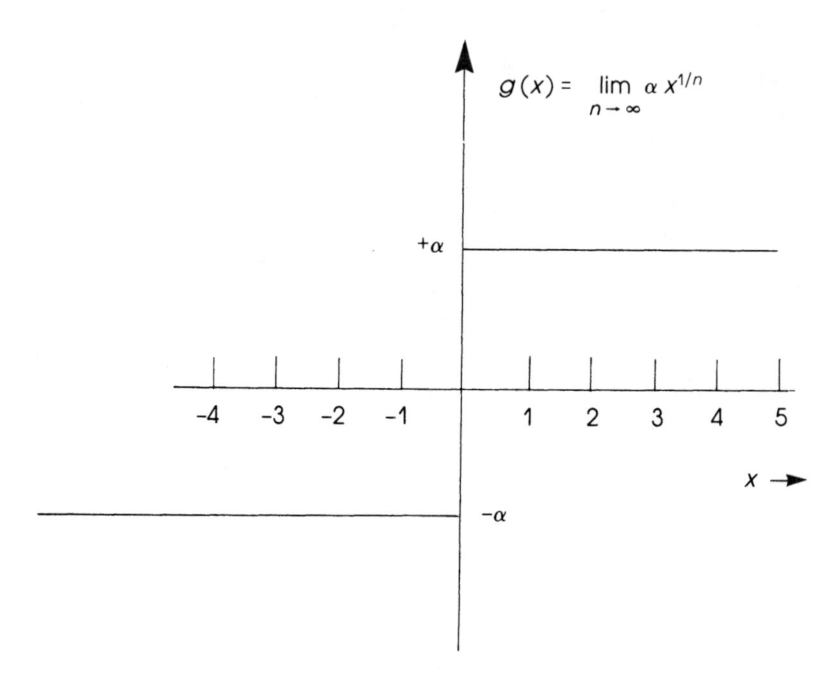

Figure 2.6 The transfer characteristic of the ideal, symmetrical limiter.

We assume that the ideal limiter is followed by a matched filter. The useful information in the limited signal is contained in the phase because all amplitude information is eliminated by the limiter action. The matched filter converts the phase information into a peaked signal with an amplitude proportional to the signal energy. At the same time the filter narrows the passband and restores amplitude fluctuation to the limited noise. The desired signal phase information can be in the form of phase coherence in a single long pulse or multiple pulses, binary-coded phase pulses, linear frequency modulation, and so on. Davenport [9] describes the effect of an ideal limiter on an unmodulated sine wave plus a Gaussian noise. If we view the limited signal and noise before filtering, the resultant waveform appears to contain no information; however, analysis shows that at most there is only a small change in J/S at the output. If the signal is much stronger than the noise jamming, the jamming is suppressed by 3 dB. If the jamming is much stronger than the signal, the noise jamming is enhanced by $4/\pi$ ($\approx$1 dB).

For a nonlinear bandpass amplifier, the effect on output J/S is nearly negligible. Davenport and Root [7, p. 310] show that for either small J/S or for large J/S:

$$(J/S)_o \approx (J/S)_i \qquad \text{for } (J/S)_i \gg 1 \text{ or } (J/S)_i \ll 1 \tag{2.4}$$

On the other hand, the nonlinear detector has the approximate characteristic:

$$\begin{aligned} (J/S)_o &\approx (J/S)_i^2 \qquad (J/S)_i \gg 1 \\ (J/S)_o &\approx (J/S)_i \qquad (J/S)_i \ll 1 \end{aligned} \tag{2.5}$$

For a bandpass limiter, we have

$$\begin{aligned} (J/S)_o &\approx \frac{4}{\pi}(J/S)_i \qquad (J/S)_i \gg 1 \\ (J/S)_o &\approx \tfrac{1}{2}(J/S)_i \qquad (J/S)_i \ll 1 \end{aligned} \tag{2.6}$$

Therefore, if the radar to be jammed by noise contains nonlinear circuits, we can take their effects into account, at least approximately, through use of Figure 2.5 or the foregoing approximations.

2.5 TRACKING RADARS AND NOISE JAMMING

Jamming tracking radars might have several purposes, including [2, p. 486]

1. Preventing acquisition of the target;
2. Delaying acquisition;
3. Preventing the radar from obtaining range or velocity data on the target;
4. Introducing large range-tracking or velocity–tracking errors;
5. Breaking lock in range or velocity;
6. Introducing errors or breaking lock in angle tracking;
7. Introducing false targets.

Subsequent chapters cover deceptive jamming of tracking systems. Here, we are concerned with the effects of noise jamming and, in particular, the vulnerability of the radar as a function of J/S. (Angle tracking of the jammer itself is often possible and is even made more accurate at high J/S.)

Preventing acquisition of a tracking radar is similar to the situation of a search radar, and the vulnerability to jamming (i.e., the J/S needed to disrupt the radar) is similar. Usually, the tracking radar has been provided with the approximate coordinates of the target; thus, it may be able to set up range and velocity gates that eliminate a significant part of the jamming noise. In this case, J/S at the output of the tracking radar receiver may be lower than that of a corresponding search radar. Naturally, if the jamming comes from the target (self-screening), we can use the jammer as a beacon to aid the acquisition process; at least for angle tracking. *J/S* values in excess of 10 dB in the IF should be sufficient to deny acquisition in range or velocity, just as it is sufficient to obliterate target echos when a threshold detector is used.

Effects of Noise on Tracking Circuits

The outputs of the radar tracking circuits provide angle, range, and velocity measurements on the target. (We can determine velocity directly from doppler frequency shift or by differentiating range data.) Tracking is usually performed by feedback circuits that operate to minimize an error signal; however, manual tracking remains an important alternative because many ECM techniques are relatively ineffective when there is a person in the loop. (Feedback tracking is discussed in Appendix A.)

A simplified radar tracker block diagram is shown in Figure 2.7. After initial filtering by the receiver, a matched filter can maximize the peak *J/S*. Ideally, the matched filter provides coherent integration of the signal before detection and tracking; however, some combination of coherent and noncoherent filtering is customary radar design practice. The matched-filter output can also be used for display and threshold detection.

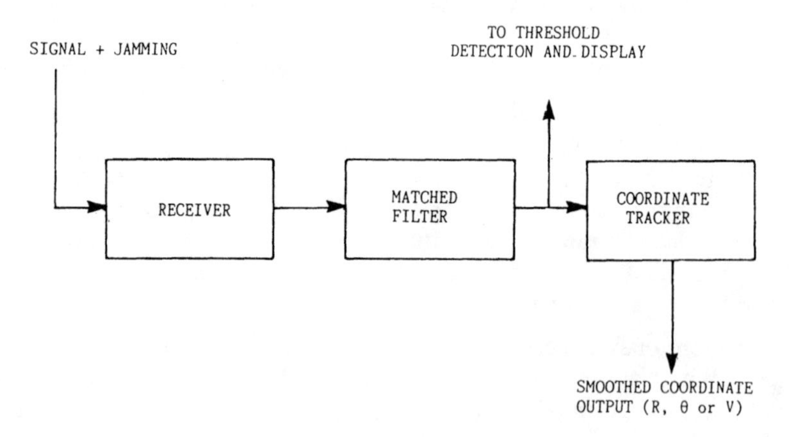

Figure 2.7 Simplified radar receiver and coordinate tracker.

The coordinate tracker differentiates the signal voltage with respect to that coordinate to establish a zero crossing at the peak of the signal. The use of the matched filter ahead of the differentiating operation ensures that the zero crossing will have maximum slope. The matched-filter output is used to normalize the differentiated signal. (This normalization can also be done by the AGC or *instantaneous* AGC (IAGC).) The operation of the AGC circuits, which also use feedback, is discussed in Chapter 6. The detected-target signal voltage, $|\Sigma|$, along with the derivative of the same signal, Δ, is shown in Figure 2.8.

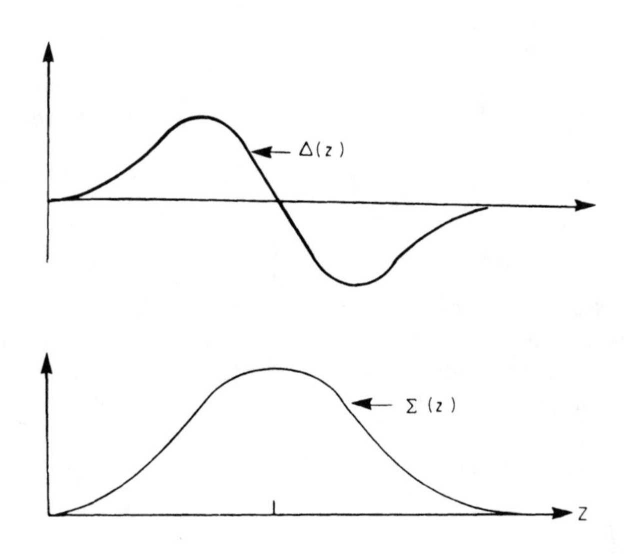

Figure 2.8 A signal (Σ) and its derivative (Δ).

The derivative signal Δ is normalized (divided) by the sum signal $|\Sigma|$ to give a slope $\Delta/|\Sigma|$ that is independent of range, target size, and other amplitude effects. A constant slope is desirable because the feedback loop gain is proportional to the slope. A reduced gain may lead to instability and a reduced servo bandwidth. With a reduced bandwidth the servo loop has less ability to follow target accelerations.

The same generic tracker description applies to all measurement coordinates (angle, range, and velocity). Although tracking can be open loop, at least over a limited region of a particular coordinate, there is an advantage in using a closed-loop tracker at $J/S > 1$. With a closed loop, tracking is about a null and second-order noise effects are minimized.

In a tracking loop, the input signal consists of the desired target signal plus jamming. For $J/S > 1$, the quality of tracking will decrease, and, for some combination of high J/S and target motion, tracking will be lost and we must repeat the acquisition phase. Barton [1] says the following about loss of (angle) track: "In practice, the radar will usually lose track as a result of target motion, servo unbalance or low-frequency oscillations (induced by the reduced gain of the servo) before the S/N ratio reaches -10 dB." We usually measure S/N in the IF amplifier before the detector, and then account for the small-signal suppression effect and integration in the narrow-band servo.

The functional relationship of the error measurement circuit is

$$\frac{\Delta}{|\Sigma|} = \frac{1}{|\Sigma|} \frac{\partial |\Sigma|}{\partial z} \tag{2.7}$$

Assume that the signal $|\Sigma|$ has a Gaussian shape when there is no jamming:

$$|\Sigma| = A\,\mathrm{e}^{-z^2/2\sigma^2} \tag{2.8}$$

where

A = peak amplitude of the signal;
σ = constant related to the width of the signal (e.g., range-gate width);
z = displacement of the present signal peak from the present position of the tracking gate.

The derivative with respect to z is

$$\Delta = A\left(\frac{-z}{\sigma^2}\right)\mathrm{e}^{-z^2/2\sigma^2} \tag{2.9}$$

Then the normalized error signal is

$$\frac{\Delta}{|\Sigma|} = -\frac{z}{\sigma^2} \tag{2.10}$$

Examination of (2.10) shows that the output has the desired characteristics—that is, a null at the current target location ($z = 0$), linear variation of the error signal with the target displacement (z), and no dependence on signal amplitude.

If the radar uses monopulse for tracking in angle, the derivative of the received signal with respect to the angle coordinates will not be affected by noise because the energy received across all parts of the aperture from the jammer is evaluated at the same time. If the angle tracker performs some kind of sequential lobing, then, because the noise amplitude changes randomly from time to time, the tracking circuitry will receive random error signals. The same is true for range and velocity trackers. For range tracking of the split-gate type, we subtract the noise amplitude averaged over one gate width from the noise amplitude averaged over the next gate width. Because the gate width is usually matched to the IF filter bandwidth of the radar, the noise in one range gate is nearly independent of the noise in the adjacent range gate. The difference of these values is also a random variable. For velocity tracking, we subtract the power in adjacent doppler filters. Again, the noise in each of these spectral bands should be independent, and the difference of the power received in adjacent filters is a random variable.

In each case, the introduction of a single random error at the input to the coordinate tracker (even a large error) does not necessarily produce a large shift in the output value of the coordinate being tracked. Because the time constants used in the tracker's feedback loops are selected to match the maneuvering characteristics of expected targets, it will usually take errors introduced over several radar measurement intervals to shift the tracker output significantly. If noise jamming is introduced at the tracker input, the error signal is just as likely to call for a shift in either direction. It requires smart-noise or deceptive jamming to move the tracker consistently in the same direction on successive measurement intervals.

Suppose the tracker is locked on to a target, but then high-level noise jamming is introduced. The resulting error signal will be randomly positive or negative, and the output of the tracker will wander (slightly and somewhat aimlessly) about the last target-tracking coordinate position. If the target coordinate being tracked is constant, the noise jamming would not immediately cause any large error in the tracking. If the coordinate is changing due to relative motion between the tracker platform and the target, the high-level noise jamming will soon create a large error because the tracker's aimless wandering will not be following the target motion at all.

Suppose each measurement is corrupted by noise such that the error signal is a uniformly distributed random variable between the limits of $\pm 0.5\alpha$. After many such disturbances, the output of the coordinate tracker will have changed by an amount that consists of a weighted average of the inputs; furthermore, the weighting function is the impulse response of the tracking loop. The output of the loop is given by

$$z_2(k) = \sum_{i=-\infty}^{k} z_1(i)h(k - i) \tag{2.11}$$

where

z_2 = tracking output;
z_1 = input to the tracker;
h = closed-loop impulse response.

For our purposes, the closed-loop response can be approximated by an average of n measurements, where n is the number of radar measurement intervals that occur during the main part of the impulse response. (If the noise bandwidth of the loop is β_n, the major portion of the impulse response of most tracking loops is contained within about $1/\beta_n$.) If the radar measurement interval is T, then the effective number of measurements averaged by the loop is about

$$n \approx \frac{1}{\beta_n T} \tag{2.12}$$

The average of n identically distributed random variables with zero mean tends toward a normal distribution with zero mean and standard deviation of

$$\sigma_n = \frac{\sigma_1}{\sqrt{n}}$$

where

σ_n = standard deviation of the tracker output after averaging n inputs;
σ_1 = standard deviation of the random error at the tracker input;
n = effective number of noise-corrupted measurements averaged by the tracking loop.

For a uniformly distributed random variable over $(-0.5\alpha, +0.5\alpha)$, the value of σ_1 is $0.5\alpha/\sqrt{3}$. At the output of the loop the standard deviation after averaging n measurements will be

$$\sigma_n = \frac{0.5\alpha}{\sqrt{3n}} \tag{2.13}$$

Because the input error can be at most 1 gate width after one noise-corrupted input, the maximum value of α is 2; that is, the error can extend from -1 gate width to $+1$ gate width. At the output of the loop, the tracker position will tend to remain as it was when the signal was last present, with a normally distributed random displacement having a standard deviation of about

$$\sigma_n \approx \frac{1}{\sqrt{3n}} \tag{2.14}$$

For large n, it is clear that the output standard deviation σ_n will be a small fraction of a gate width. This means that most well-designed tracking loops will be "frozen" by high-level noise jamming, and will continue to "coast" along the path established for the target before the onset of the jamming. This means that if the target maneuvers after the onset of the jamming, tracking will cease and the radar will need to reacquire the target. It also means that noise, by itself, will probably not displace the tracker off the target coordinate. Only motion by the target or drift in the tracking loops creates significant tracking errors (in excess of 1 gate width).

The effect of the tracker on J/S is the same as noncoherent integration. The tracking loop integrates (averages) several measurements, and the J/S value at the output of the tracker is reduced just as it would be by a noncoherent integrator. The main difference is that the closed-loop tracker adjusts, based on the result of the integration. Thus, if J/S at the input to the tracker is 10 dB, and if the tracker impulse response averages 100 measurements, J/S at the tracker output will be reduced by a factor of about 0.1, reducing the J/S at the output to 0 dB. We will estimate tracker performance at low J/S values by using Barton's methods (discussed in the next section). At high output J/S values, the tracking action is essentially independent of the current target coordinates.

The effect of the tracking loop is to perform postdetection integration. Hence, the loop can continue to track even when S/N at the input is below 0 dB. Then the problem is how to determine that the loop is still tracking—in other words, if there is no indication of a signal, what is there to give confidence to the operator that the object being tracked is still the target? The usual case is that the target disappears from the operator's display before the tracking action ceases.

2.6 TRACKING AT LOW S/N

Barton [1, p. 368; 2, p. 467] discusses the action of tracking loops operating at a low S/N. In general, the rms error due to noise is given by expressions of the form:

$$\sigma_z = \frac{\Delta z}{\sqrt{(S/N)n}} \tag{2.15}$$

where

σ_z = rms error in tracking of coordinate z;
Δz = effective gate width in coordinate z (e. g., related to antenna beamwidth for angle tracking, range-gate width for range tracking, and doppler filter bandwidth for velocity tracking);
S/N = signal-to-noise ratio at tracker input (note that any losses in the detection process must have been taken into account);
n = equivalent number of coordinate measurements integrated by the tracking loop (e.g., PRF/β_n, where PRF is the radar's pulse repetition frequency and β_n is the bandwidth of the tracking loop).

This equation is valid so long as the quantity under the radical in the denominator is greater than about 7. Since n can be rather large, it is possible to maintain the product $(S/N)n > 7$ even if $S/N < 1$.

For noise jamming, J/S at the tracker output is approximately

$$\left(\frac{J+N}{S}\right)_{\text{out}} \approx \left(\frac{J+N}{S}\right)_{\text{in}} \frac{1}{\sqrt{n}} \tag{2.16}$$

If this is less than 1/7, then the tracker will track the target with random errors due to jamming and noise that approach Δz. If the quantity in (2.16) is greater than about 10, tracking ceases and the loop "coasts." For output $(J + N)/S$ between 1/7 and 10, tracking is possible at times, but the target is easily lost if there is high acceleration, drift in the circuitry, or other disturbances to the system.

The radar designer must choose the servo response characteristics with care. The response must be fast enough to follow targets of interest but slow enough to permit averaging several measurements to improve accuracy and reduce vulnerability to noise jamming. Barton [1,2] suggests that the optimal servo bandwidth is that which minimizes the sum of the squared errors due to lag, noise, and any other factors of importance. Considering the lag and noise as the two key contributors, it is clear that the lag in following target motion is reduced as the loop bandwidth is increased. However, a wider servo bandwidth allows more noise jamming to reach the output. The relationship between servo noise bandwidth and equivalent averaging time of the loop is [2, p. 464; e.g., 10.2.3]

$$\beta_n \approx \frac{1}{2t_0} \tag{2.17}$$

Because the position of a target that has acceleration a in coordinate z is changing as $\frac{1}{2}at^2$, and because the servo loop output is approximately the average position during time T, the tracking lag error of such a loop is approximately

$$\epsilon_{\text{lag}} = \frac{at_0^2}{6} = \frac{a}{24\beta_n^2} \tag{2.18}$$

The rms error due to noise, from (2.15), is

$$\sigma_z = \Delta z \sqrt{\frac{\beta_n T}{S/N}} \tag{2.19}$$

The sum of the squared errors is then

$$\epsilon^2 = \epsilon_{lag}^2 + \sigma_z^2 = \frac{a^2}{(24)^2\beta_n^4} + \frac{(\Delta z)^2\beta_n T}{S/N} \tag{2.20}$$

This gives the optimal bandwidth as

$$\beta_{n0} = \left[\frac{a^2(S/N)}{144(\Delta z)^2 T}\right]^{1/5} \tag{2.21}$$

Choosing typical radar and target values gives a typical range tracker servo bandwidth:

$a \approx 8$ g ≈ 80 m/s^2;
$\Delta z = 75$ m;
$T =$ PRI $= 10^{-3}$ s;
$S/N = 10$;
$\beta_{n0} \approx 2.4$ Hz;
$t_0 \approx \frac{1}{2}\beta_{n0} \approx 0.208$ s;
$n \approx 208$ pulses.

For $n = 208$, J/S at the output of the tracker is reduced by 5 log208 = 11.6 dB compared with J/S at the input. Hence, to obliterate tracking performance, we want the value of J/S at the tracker input to be about 11.6 + 10 = 21.6 dB.

2.7 A TYPE 1 SERVO EXAMPLE

The closed-loop response of the tracking system of Figure 2.9 is

$$Z_2(s) = \frac{Z_1(s)}{s^2/\omega_n^2 + (2\zeta/\omega_n)s + 1} \tag{2.22}$$

The impulse response is the response when $Z_1(s)$ is a constant ($z_1(t) = \delta(t)$), which is

$$z_2(\tau) = \frac{\omega_n}{\sqrt{1-\zeta^2}}\,[e^{-\zeta\omega_n t}]\,\sin\,[\omega_n\sqrt{1-\zeta^2}\,\tau] \tag{2.23}$$

The impulse and step responses of this servo (for $\zeta = 0.5$) are shown in Figure 2.10. The normalized time axis is in units of $\omega_n t$. As can be seen, the major part of the impulse response is about 2 units wide; hence, the output consists (approximately) of the average of the radar measurements during this time.

The overshoot in the step response is a consequence of underdamping ($\zeta < 1$). Figure 2.11 indicates the response of this servo to a ramp input (i.e., constant-velocity target in coordinate z) and the lag in the servo response. Figure 2.12 shows the same response when the servo bandwidth is doubled ($\omega_n = 2$); this reduces the lag error by a factor of 2. (The width of the impulse response is reduced by a factor of 2.)

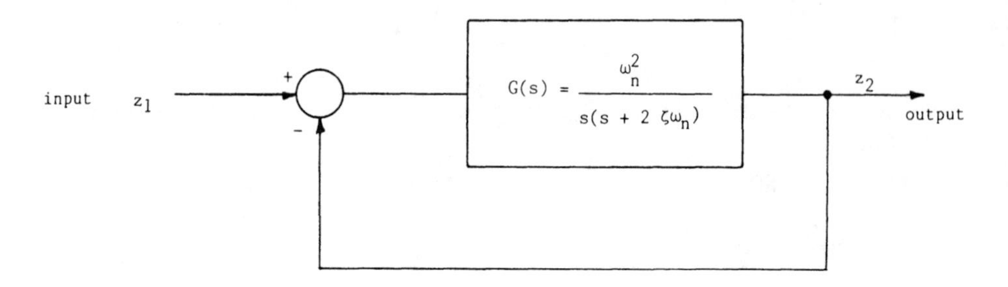

Figure 2.9 Type 1 servo block diagram.

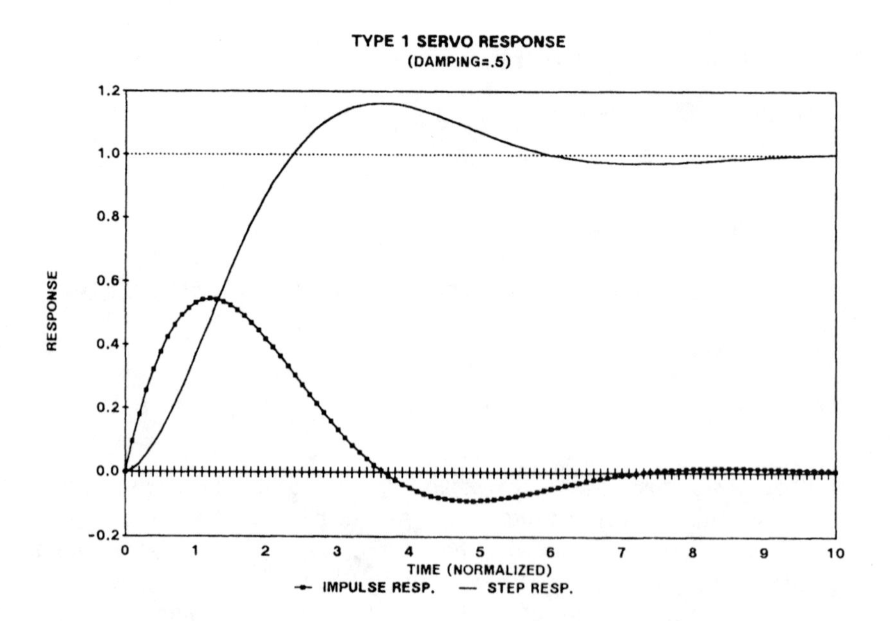

Figure 2.10 Response of a type 1 servo.

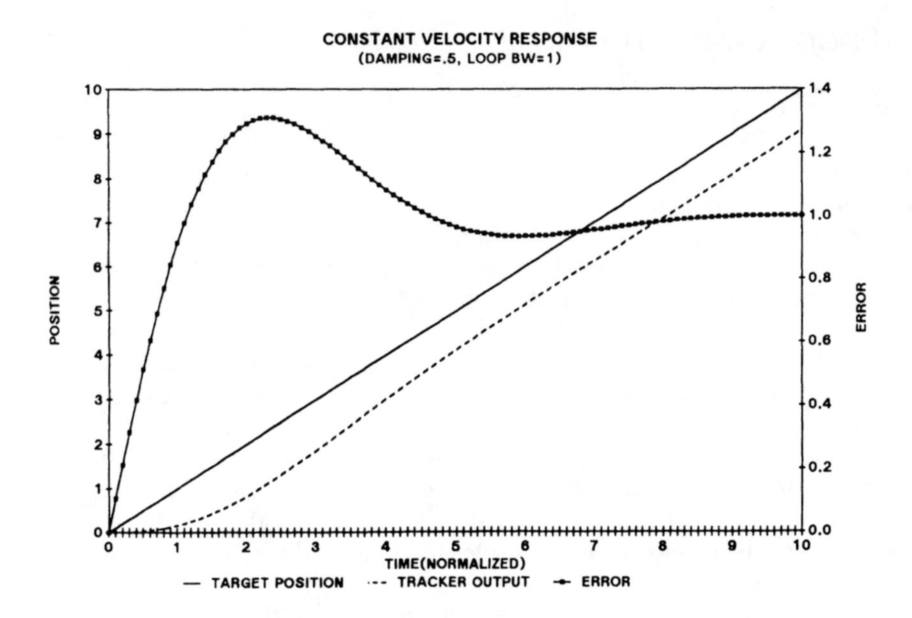

Figure 2.11 Lag error of type 1 servo ($\omega_n = 1$).

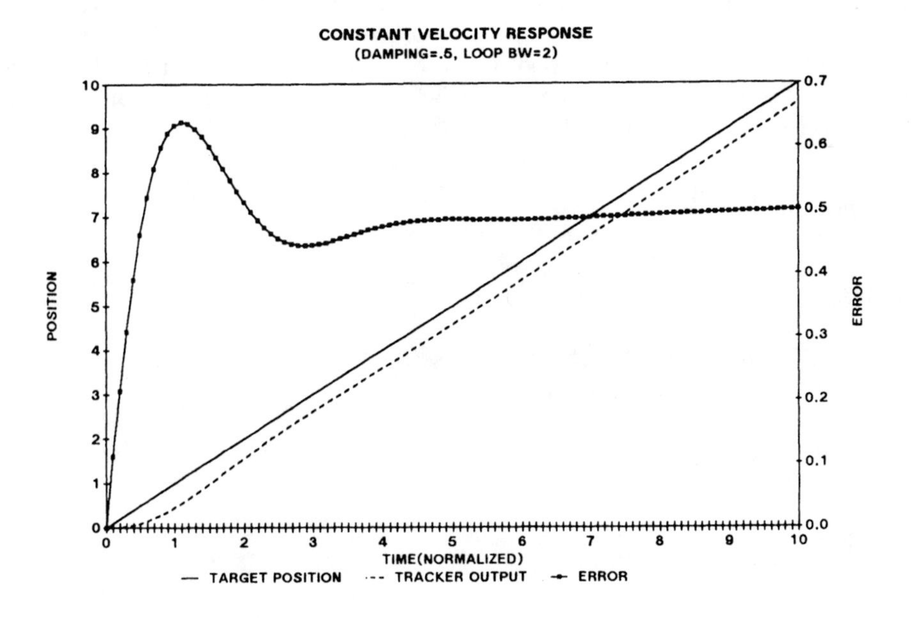

Figure 2.12 Lag error of type 2 servo ($\omega_n = 2$).

2.8 NOISE GENERATION

The generation of high-level noise for jamming purposes can be accomplished in several ways. The quality of the noise is best if it has a constant spectral density across the band of interest and if it has a normally distributed amplitude. Of course, any sample of finite duration taken from such a random process can be judged only on the basis of the properties of that sample, which in general will not have a flat frequency spectrum and normal amplitude distribution. Noise having a constant power spectral density over a restricted band is often called *band-limited white noise.* We can generate this type of noise with high quality by direct noise amplification, using a noise source and a power amplifier that covers the band of interest. Such noise might typically be 10 to 100 MHz in bandwidth, and it can be "tuned" to a given center frequency by mixing it with an appropriate carrier (sinusoid). The modulation process can be single- or double-sideband *amplitude modulation* (AM), in which case part of the carrier often leaks through to the output. This reduces the available noise-jamming level somewhat and can serve to identify the noise source as a jammer. The bandwidth of each sideband that results is the same as that of the noise used to modulate the carrier.

If the noise is used to frequency-modulate a carrier, the resulting signal is also noiselike and will be centered at the frequency of the carrier. In this case, the bandwidth of the resulting jamming is a function of both the bandwidth and the amplitude of the noise modulation. As shown in Figure 2.13, the noise voltage $n(t)$ is used to frequency-modulate a *voltage-controlled oscillator* (VCO). The frequency of the VCO is assumed to be proportional to the voltage applied or $f = k_f v$. Suppose the noise source is white and Gaussian. Then the probability that the noise voltage is between n and $n + \mathrm{d}n$ is

$$p(n)\,\mathrm{d}n = \frac{1}{\sigma\sqrt{2\pi}}\,\mathrm{e}^{-n^2/2\sigma^2}\,\mathrm{d}n \tag{2.24}$$

where

$p(n)$ = probability density of the noise;
σ = standard deviation of the noise.

Note that σ is related to the noise power by

$$\frac{\sigma^2}{R} = kTB \tag{2.25}$$

where

R = resistance of the load at the VCO input (e.g., 50 Ω);
k = Boltzmann's constant;
T = noise temperature of the noise source.

Then the probability that the output instantaneous frequency is between f and $f + \mathrm{d}f$ is

$$p(f)\,\mathrm{d}f = \frac{1}{\sqrt{2\pi}k_f\sigma}\,\mathrm{e}^{-f^2/2k_f^2\sigma^2}\,\mathrm{d}f \tag{2.26}$$

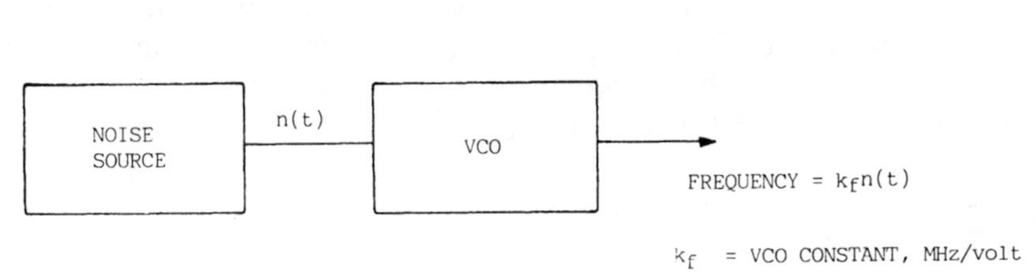

Figure 2.13 FM noise generator.

Suppose $k_f\sigma \ll B$. Then the amount of frequency deviation ($k_f\sigma$) will nearly always be small compared with the bandwidth (B) of the modulation. This is narrow-band *frequency modulation* (FM), and the spectrum at the VCO output is approximately $2B$ in width and is concentrated near the carrier. We can approximate it by dividing the modulating signal into narrow-band segments of bandwidth δB. Each of these components will be nearly sinusoidal and will contain a fraction ($\delta B/B$) of the noise power. Each such nearly sinusoidal component will therefore have a peak amplitude of $[2kT\delta BR]^{1/2}$ and will cause a peak frequency deviation of about

$$\Delta F = k_f[2kT\delta BR]^{1/2} \tag{2.27}$$

For narrow-band FM, the spectrum of the modulated wave has the form:

$$S(f) \approx \begin{cases} \left(\dfrac{\Delta F}{2(f - f_c)}\right)^2, & \left(f_c - \dfrac{B}{2}\right) < f < \left(f_c + \dfrac{B}{2}\right), \quad f \neq f_c \\ \delta(f), & f = f_c \end{cases} \tag{2.28}$$

This spectrum is not white, but falls off in proportion to the reciprocal of the square of the separation of the frequency of interest and the carrier frequency f_c.

Suppose $k_f\sigma > B$. Then the amount of frequency deviation ($k_f\sigma$) is large compared with the bandwidth (B) of the modulating noise. This is wideband FM, and therefore the spectrum of the frequency-modulated output is approximately the same shape as the probability density function of the input [10]. This means that the spectrum is approximately Gaussian with "standard deviation" $k_f\sigma > B$. We take the bandwidth of the output spectrum as $k_f\sigma$. Then it is clear that we can make this bandwidth arbitrarily large by making the noise voltage at the VCO input larger and larger. However, the resulting signal is *not* high-quality noise. The VCO may be viewed as a frequency-hopped signal that only rarely "visits" a particular frequency. Although there is power at each frequency in the band, it has an impulsive nature and lacks quality. To get high-quality noise using FM, we want the amount of frequency deviation at the output and the bandwidth of the noise modulation at the input to be about the same:

$$k_f\sigma \approx B \tag{2.29}$$

REFERENCES

1. Barton, D.K., *Radar System Analysis,* Artech House, Norwood, MA, 1976.
2. Barton, D.K., *Modern Radar System Analysis,* Artech House, Norwood, MA, 1988.
3. Skolnik, M.I., ed., *Radar Handbook,* 1st Ed., McGraw-Hill, New York, 1970.
4. Meyer, D.D., and H. A. Mayer, *Radar Target Detection,* Academic Press, New York, 1973.
5. Wiley, R.G., *Electronic Intelligence: The Interception of Radar Signals,* Artech House, Norwood, MA, 1985.
6. Wiley, R.G., *Electronic Intelligence: The Analysis of Radar Signals,* Artech House, Norwood, MA, 1982.
7. Davenport, W.B., Jr., and W. L. Root, *An Introduction to the Theory of Random Signals and Noise,* IEEE Press, 1987 (republication of the 1958 McGraw-Hill edition).
8. Lawson, J.L., and G. E. Uhlenbeck, *Threshold Signals,* M.I.T. Radiation Laboratory Series, Vol. 24, McGraw-Hill, New York, 1950.
9. Davenport, W.B., Jr., "Signal-to-Noise Ratios in Band-Pass Limiters," *Journal of Applied Physics,* Vol. 24, No. 6, June 1953, pp. 720–727.
10. Blachman, N.M., and G. A. McAlpine, "The Spectrum of a High-Index FM Waveform: Woodward's Theorem Revisited," *IEEE Transactions on Communications Technology,* Vol. COM-7, No. 2, April 1969.

Chapter 3
RANGE DECEPTION

3.1 RANGE-TRACKING LOOP DESCRIPTION

In a radar-controlled weapons system, range tracking serves two essential functions: (1) It provides the value of the target range, R_T, for solution of the fire control or weapon guidance problem; and (2) it provides a target acceptance gate (range gate) that excludes clutter and interference from other ranges. The second function is essential to the proper operation of the other target coordinate tracking loops. If an enemy jammer succeeds in disrupting the radar's range-tracking loop, the target acceptance gate loses the target and is no longer able to pass along the target signal to the angle trackers or to the doppler tracker, if it exists (Figure 3.1). Jamming specifically aimed at luring the range tracker off the target is called *range-deception ECM*.

In a monostatic pulsed radar, range measurement amounts to measurement of the round-trip time delay, t_T, of the echo pulse and calculation of the target range, R_T, from the equation:

$$R_T = \frac{c}{2} t_T$$

where c is the velocity of light. Figure 3.2 shows the essentials of a range-tracking loop. The tracker output, $\hat{t}_T$, is the tracker's estimate of target range delay t_T. The measurement error

$$\epsilon_t = t_T - \hat{t}_T$$

exists within the time discriminator with output voltage:

$$V_\epsilon = g_t(\epsilon_t)$$

For linear loop operation, g_t should be a linear function of ϵ_t.

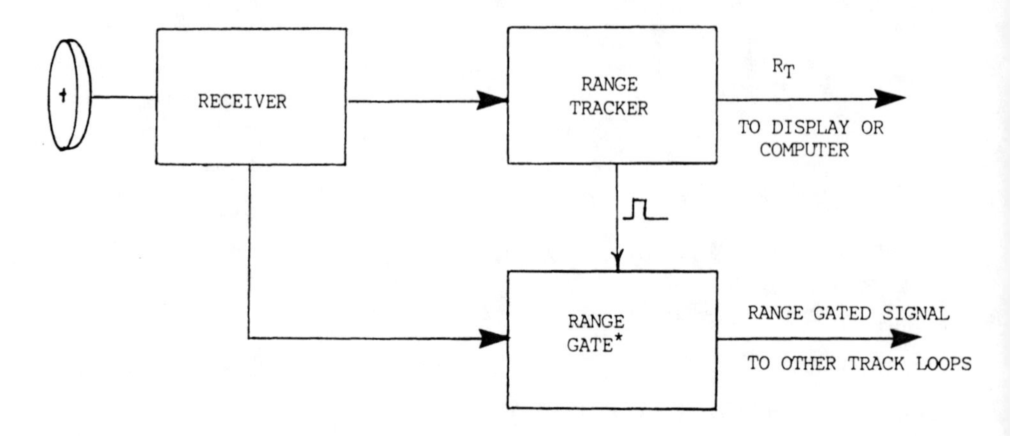

*Gate duration typically equal to pulse duration.

Figure 3.1 Functions of range tracker.

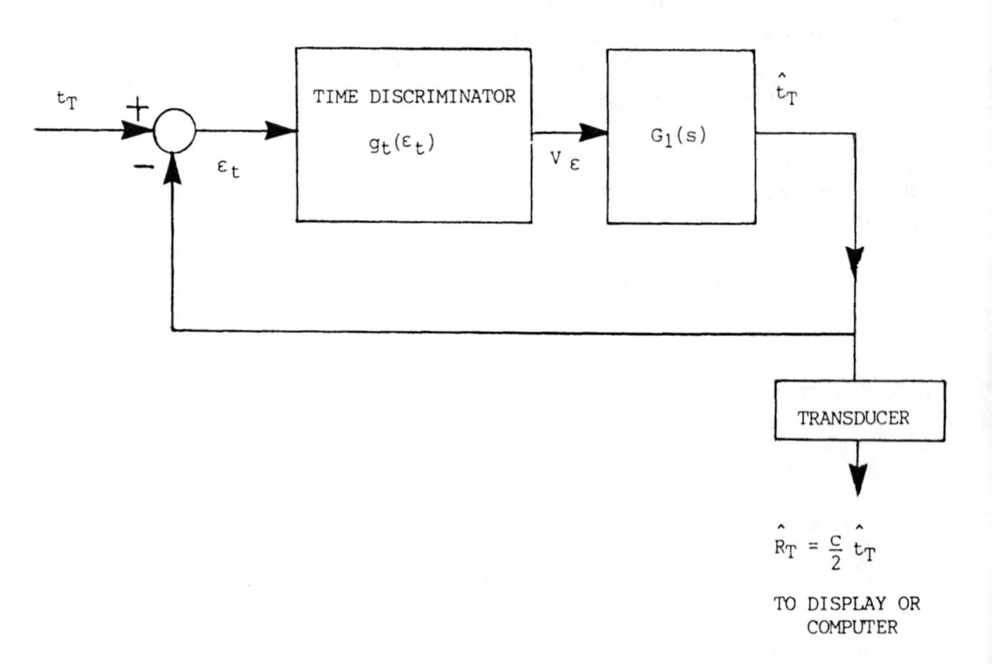

Figure 3.2 Range tracker block diagram.

3.2 SPLIT-GATE TIME DISCRIMINATOR

The most prevalent form of time discriminator is the split-gate discriminator (Figure 3.3). In Figure 3.2, $\hat{t}_T$ represents the time location (relative to the transmitted pulse time t_0) of the center of the split gate. When the split gate is positioned so that the early and late gates capture equal shares of the target pulse, the difference of the two gate outputs is zero, and the gate can be said to be centered on the target. The acceptance gate (range gate of Figure 3.1) is slaved to the split gate; thus, when the split gate is centered on the target pulse, the target pulse falls within the range gate and is passed along for processing by the other tracking loops. The split-gate tracker is often called a *centroid* tracker. This is a slight misnomer, for the tracker strives to balance the areas of the target pulse waveform falling in the early and late gates rather than balancing the first moments of the waveform. Usually, this point is rather trivial; for symmetrical pulses, the equal-area point is the same as the centroid. If $p(t)$ is the target waveform (Fig. 3.3), the split-gate discriminator output is

$$V_\epsilon(t) = \int_{\text{early gate}} p(t)\,dt - \int_{\text{late gate}} p(t)\,dt \tag{3.1}$$

Evidently, for a target pulse of the sort in Figure 3.3, V_ϵ will be, for small errors, an approximately linear function of the error $t_T - \hat{t}_T$.

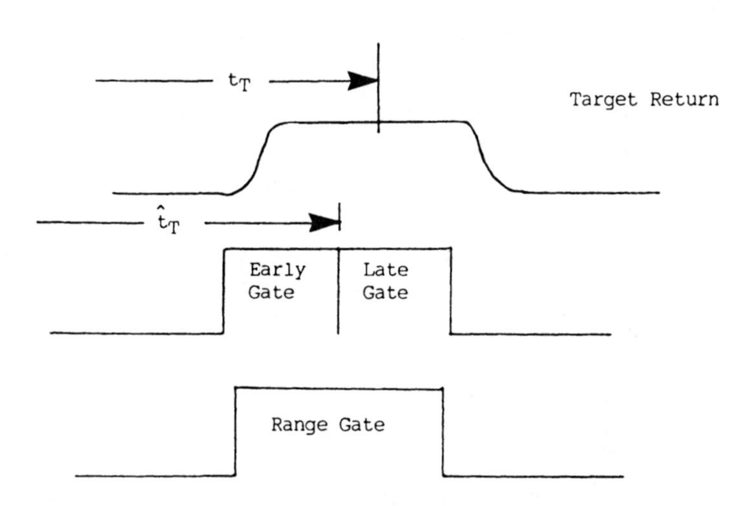

Figure 3.3 Split-gate time discriminator.

A simple and effective way of deriving the early- and late-gate integrals is to apply the signal $p(t)$ to an RC quasi-integrator (Figure 3.4). The capacitor is discharged before the arrival of the next expected signal pulse. For a rectangular signal pulse $p(t) = V_1$, which completely overlaps the gate closure interval t_g, the capacitor voltage reaches the value:

$$V_c = V_1[1 - e^{-t_g/RC}]$$

This result can be compared with V_i, the output of the ideal integrator:

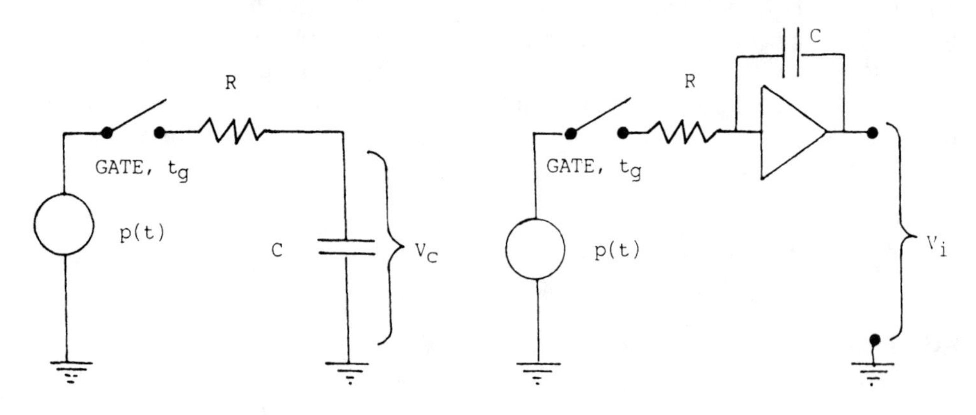

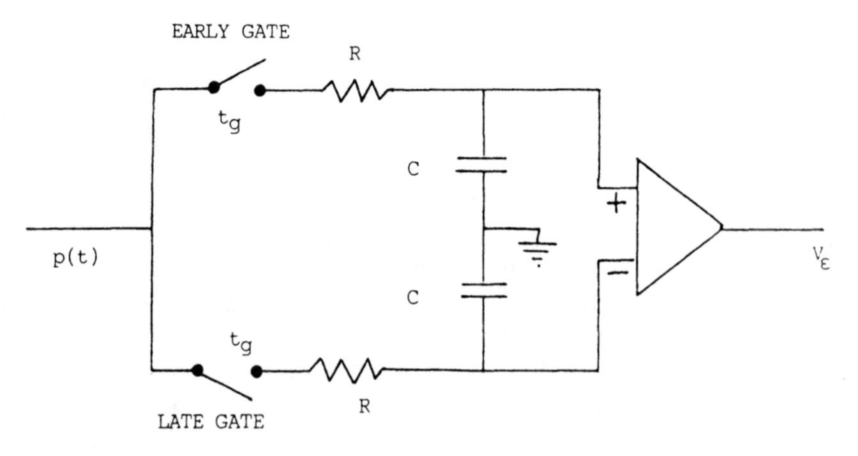

Figure 3.4 Development of split-gate time discriminator.

$$V_i = \frac{1}{RC}\int_0^{t_g} V_1\,dt = \frac{V_1 t_g}{RC}$$

The fractional error of the quasi-integrator is

$$\frac{V_c - V_i}{V_i} = \frac{V_c}{V_i} - 1 \approx -\frac{1}{2}\frac{t_g}{RC}$$

To achieve an error of less than 1% requires

$$RC > 50t_g \tag{3.2}$$

Figure 3.4(c) shows a configuration for subtracting the outputs of a pair of RC integrators in an operational amplifier to form the error voltage V_ϵ of (3.1). Not shown is the provision for dumping the charge of the capacitors.

3.3 SAMPLE-AND-HOLD (BOXCAR) CIRCUIT

If the gate paths (the switches in Figure 3.4) have sufficiently high leakage resistance in their "off" state, and if the input resistance of the amplifier is sufficiently high, the capacitors hold their charge and the amplifier output remains constant at V_ϵ until the capacitors are dumped. This is one way of forming a holding or *boxcar* circuit.[1] Figure 3.5 shows the stairstep replication of a continuous waveform produced by a periodic sample-and-hold operation. In a system such as a pulsed radar, in which short samples of a variable are available at times spaced by the pulse repetition interval, the boxcar circuit is often used to "stretch" the short pulse sample to a duration almost equal to the PRI. Frequently, such stretched samples are fed into a smoothing filter (e.g., the actuator elements of a servomechanism) that converts the stairstep waveform to a continuous function of time. Clearly, in this case the boxcar operation increases the low-frequency gain of the path in which it is placed. If the pulse sample of amplitude V_1 and duration τ is "smoothed" or averaged by the smoothing filter, the smoothed output may approximate the average:

$$V_{av} = V_1 \frac{\tau}{\text{PRI}}$$

which may be very small relative to V_1. If the boxcar operation is interposed, the sample is stretched to a duration almost equal to the PRI, so the average becomes approximately equal to V_1 itself.

[1]The term *boxcar* will be used in subsequent discussions to identify the sample-and-hold function.

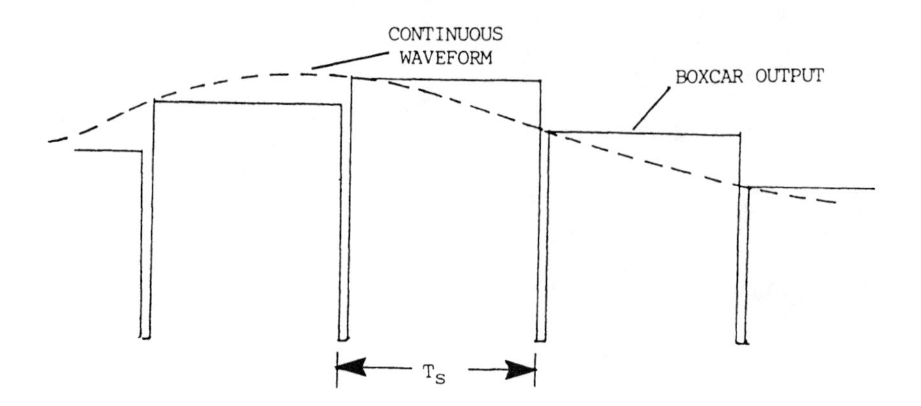

Figure 3.5 Boxcar circuit operation.

For systems in which the intersample period, T_S, is short compared with the system time constants, this gain increase achieved by sample stretching is the main effect of the boxcar. One can easily derive the boxcar transfer function $G_B(s)$. The boxcar impulse response is a unit step initiated at the sampling instant, followed by a negative unit step at the next sampling instant, T_S later. This second step represents the dumping of the previous sample. The time between the dump and the acquisition of the next sample is considered to be so short that the two operations are essentially coincident in time. The transfer function is

$$G_B(s) = \frac{1}{s} - \frac{e^{-sT_s}}{s} \tag{3.3}$$

We obtain the transfer function, $G_B\,(j\omega)$, of the boxcar circuit by setting $s = j\omega$ in (3.3):

$$G_B(j\omega) = T_s\, e^{-j\pi\omega/\omega_s} \left[\frac{\sin(\pi\omega/\omega_s)}{\pi\omega/\omega_s} \right]$$

where $\omega_s = 2\pi/T_s$. The first factor reveals the phase lag corresponding to the constant-time lag:

$$\text{time lag} = \frac{\pi}{\omega_s} = \frac{T_s}{2}$$

The magnitude of the transfer function, $|G_B(j\omega)|$, and the time lag are plotted in Figure 3.6. For an input signal that has a spectrum concentrated well below ω_s (in the fairly flat portion of the $|G_B(j\omega)|$ curve), the boxcar output tends to be a good replica of the input, delayed by $T_s/2$. Although the boxcar contributes the finite phase lag $\phi = \omega T_s/2$, this lag usually does not have a significant impact on the stability of a servo loop containing the boxcar because the sampling frequency ω_s is normally far above the loop's cut-off frequency (the frequency at which loop gain falls below unity; see Appendix A).

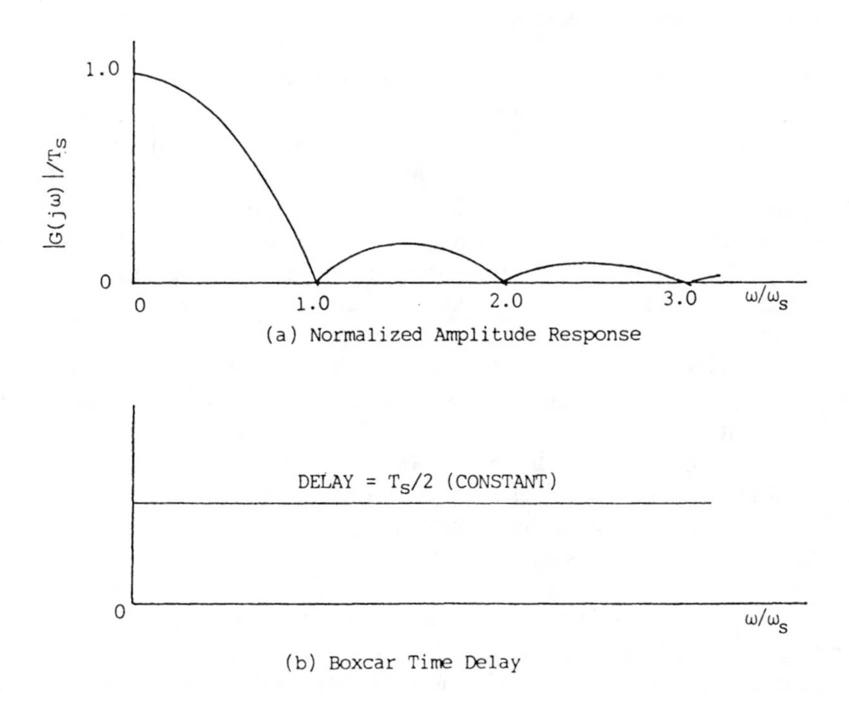

Figure 3.6 Boxcar transfer function (amplitude response and delay).

3.4 RANGE-TRACKING LOOP IMPLEMENTATION

The area-balancing feature of the split gate plus the pulse-stretching feature of the boxcar combine to form a satisfactory time discriminator that is implemented in analog circuit form in many range trackers. We can implement these functions digitally, but the area computation would require the signal pulse to be sampled at many equally spaced points.

The remainder of the range-tracking loop performs the automatic nulling of the time discriminator (error detector) output. Many different schemes are possible for its implementation. Figure 3.7 gives block diagrams for (a) an electromechanical implementation, (b) an electrical analog implementation, and (c) a digital implementation. We shall assume in each case that the time discriminator is of the form we have already discussed. The other cascaded elements combine to form the block with transfer function labeled $G_l(s)$ in Figure 3.2.

Electromechanical Implementation

Figure 3.7(a) illustrates the general configuration of the range trackers, developed during World War II and used for a long time thereafter. The filter block contains elements of the motor controller (e.g., the motor field circuit, if field control is employed) together with filter elements to shape the overall open-loop transfer function to achieve stability and a suitable transient response (see Appendix A). The motor drives a linear potentiometer up or down, according to the polarity of error voltage V_ϵ, to produce an output voltage V_r proportional to the estimated target range. Voltage V_r is compared with a linear voltage ramp that is triggered at $t = t_0$, the time of initiation of the transmitted pulse. A voltage comparator produces a range trigger at the instant at which the ramp voltage is equal to V_r, and this trigger initiates the split gate and the receiver range gate (the acceptance gate).

If the split gate is centered on the target, $V_\epsilon = 0$ and the motor remains stationary. If $V_\epsilon \neq 0$, the motor drives the potentiometer in the correct direction to shift the split gate closer to the center of the target pulse (see Figure 3.3).

The motor in the loop serves an error integration function. If motor speed is proportional to V_ϵ, then the shaft position (and V_r, which is proportional to motor shaft position) is proportional to the time integral of the error. Thus, for any steady value of target range there will be no tracking position error, because any error, however small, would eventually integrate up to a large correction voltage. The electromechanical integrator involving a motor was used in early trackers because no satisfactory electronic integrator was available. The vacuum-tube amplifiers that would have been needed for electronic integration were susceptible to dc drift. For long-term integration, the amplifiers must be direct-coupled, with the result that dc drift produces an integration error. If no other integration appears in the loop, it is called *type 1 servo.*[2] A second motor integrator was incorporated into some range trackers, producing a type 2 servo capable of following a constant-velocity target with zero lag error. Eventually, electronic integrators with low drift were developed, with the result that range trackers could be designed without the inertia and bulk of the electromechanical units.

[2]Appendix A explains servo-type classification.

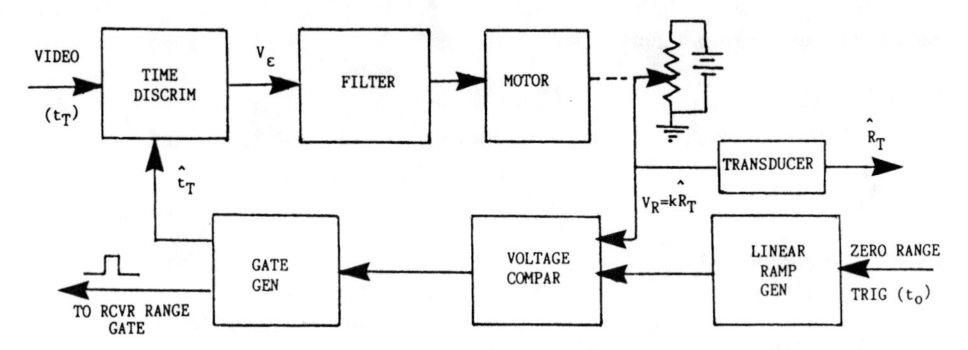

(a) Electromechanical Implementation

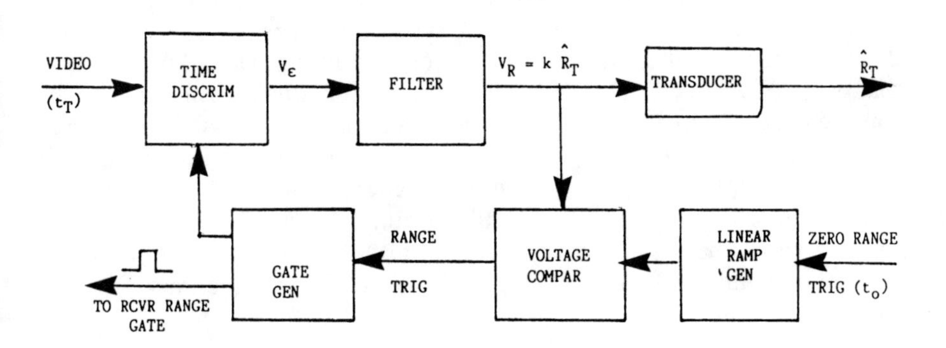

(b) Analog Electronic Implementation

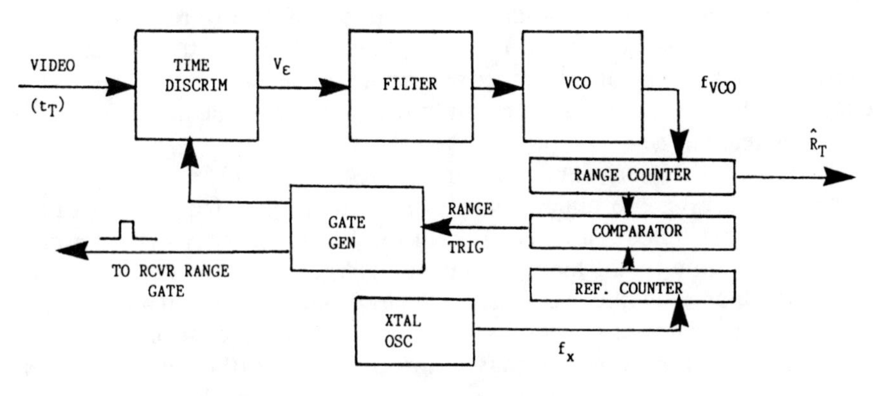

(c) Digital Implementation

Figure 3.7 Three range tracker implementations.

Electrical Analog Implementation

Figure 3.7(b) shows a tracker in which the motor integrator is replaced by an electronic integrator, contained in the filter block, to yield an all-electronic system. All other elements of the tracker are identical to those of Figure 3.7(a).

Digital Implementation

Figure 3.7(c) shows a tracker in which a part of the analog circuitry of the foregoing system is replaced by digital circuitry. The range voltage V_r is replaced by a range count, n_R, contained in the range counter register. The voltage ramp of Figure 3.7(b) is replaced by the count in the reference counter register. The range counter counts pulses generated by a VCO, for which the frequency is proportional to the filtered error voltage V_ϵ. The reference register counts pulses generated from a crystal oscillator source. The crystal oscillator and the VCO frequencies are sufficiently high that the time increment (oscillator period) amounts to a small range increment. The range count is the time integral of the VCO frequency, integrated over the counting period. The range trigger pulse is generated when the count in the reference register (counting the crystal oscillator output) matches the range count. The counting period begins at zero range time (instant of launching of the transmitted pulse) and ends at the instant of coincidence of the two counters. Thus, at any instant, the estimated range delay, $\hat{t}_T$, is

$$\hat{t}_T = n_R T_X$$

where n_R is the range count that produced coincidence of the two registers, and T_X is the period of the crystal oscillator. If $\hat{t}_T$ is slightly less than the true target delay, t_T, the error voltage will be of the proper polarity to cause an increase in the VCO frequency, thereby causing the range count to increase during the next PRI, with a consequent increase in $\hat{t}_T$.

Methods exist for interpolating the time between $n_R T_X$ and $(n_R + 1)T_X$ so that unduly high clock rates are not required to keep time quantizing errors within acceptable limits. As noted, the integration of the VCO frequency that is proportional to V_ϵ results in a type 1 loop (see Appendix A) even if no other integration is present in the loop. The filter can contain an electronic integrator if a type 2 loop is required. The presence of integrators in a tracking loop to control steady-state error yields an incidental benefit in that an integrator input (the derivative of the output) may be put to use. For instance, in a type 2 range tracker with range R, appearing at the output of the final integrator, the range rate $\dot{R}$ appears at the input to that integrator and the range acceleration $\ddot{R}$ appears at the input to the preceding integrator. In a radar with a doppler measurement capability, cross-checking between $\dot{R}$ and the target doppler can reveal range spoofing by a jammer. Unrealistically high values of measured $\dot{R}$ and $\ddot{R}$ may indicate the presence of range-deception ECM.

3.5 REALISTIC TARGET RANGE MOTION

A valuable ECCM feature of some radars is an ability to reject a false target with motion that fails to resemble the motion of a real target. Therefore, a range-deception jammer should present false targets with range rates $\dot{R}$ and accelerations $\ddot{R}$ that do not exceed the values expected of real targets of the class being mimicked. For a target capable of a maximum velocity v_{max}, the highest range rate that will be observed by a stationary radar is

$$\dot{R}_{max} = v_{max}$$

which occurs for target motion along a radial path.

The acceleration $\ddot{R}$ observed by the radar can arise from any of three sources:

1. True linear acceleration of the target along its flight path;
2. Centripetal acceleration of the target as it executes a turn at constant velocity along an arc;
3. Transformation into the radar range coordinate of target motion along a non-radial straight line at constant velocity.

The largest value of observed $\ddot{R}$ arising from true linear acceleration (1) is

$$\ddot{R}_{max} = \alpha_{max}$$

where α_{max} is the linear acceleration limit of the target, and this is observed for a radial course. The largest value of observed $\ddot{R}$ arising from centripetal acceleration (2) is

$$\ddot{R}_{max} = \left(\frac{v^2}{\rho}\right)_{max}$$

where v is the target velocity along the arc of radius ρ. This value will be observed at the instant when the radar LOS to the target is normal to the flight path. If the turn is executed at $v = v_{max}$, the centripetal acceleration is limited by the shortest radius, ρ_{min}, that can be achieved. The largest value of observed $\ddot{R}$ arising from the straight-path, constant-velocity situation (3) is

$$\ddot{R}_{max} = \frac{v_{max}^2}{R_{min}}$$

which occurs at the point of closest approach ($R = R_{min}$) as the target passes by the radar.

Example: Suppose that a radar is programmed to expect aircraft targets with the following limitations:

$v_{max} = 1000$ ft/s;

$\alpha_{max} = 2g = 64$ ft/s^2;

$(v^2/\rho)_{max} = 3g = 96$ ft/s^2 (maneuver limit).[3]

The radar might then reject any target with observed $\dot{R}$ above 1000 ft/s or with $\ddot{R}$ in excess of plausible limits. The apparent acceleration, v^2_{max}/R_{min}, can greatly exceed the other two limits when the aircraft passes at short range. For a pass at a range of $R_{min} = 1$ nmi $= 6080$ feet, the observed $\ddot{R}$ at $v_{max} = 1000$ ft/s is

$$\ddot{R}_{max} = \frac{10^6}{6080} = 164 \text{ ft/s}^2 \approx 5g$$

Thus, to make an intelligent decision on the $\ddot{R}$ acceptance limit, the radar should use available knowledge concerning the target's flight path. For a moving radar (e.g., an airborne radar) to make such decisions, it would have to subtract from observed $\dot{R}$ and $\ddot{R}$ values the contributions arising from its own platform motion.

3.6 RANGE-GATE PULL-OFF DESCRIPTION

Range-deception ECM against a pulsed radar is called *range-gate pull-off* (RGPO) and other, similar names. It is a self-protection tactic, typically employed by an aircraft or a missile with the intent of dislodging the enemy's range tracker. The ultimate goal is to cause the tracker to completely lose the target (this is called *break lock*) or, failing that, to create large enough range measurement errors to degrade the accuracy of the enemy fire control system.

RGPO is initiated by causing a jammer aboard the target vehicle to emit a false-target pulse that arrives at the radar as nearly as possible in time coincidence with its own target-echo (skin-return) pulse. If the false-target pulse power appreciably exceeds the power of the skin-return pulse, it tends to dominate the tracker. The false-target pulse can then be delayed successively, more and more, with the intent of luring the gates of the range tracker away from the skin return. Once the range tracker is pulled completely off the target, the J/S within the range-gated receiver is infinite because the target skin return is no longer present in the range gate. Other tactics, such as angle or doppler deception, may then be initiated to

[3] g = acceleration of gravity ≈ 32 ft/s^2

disrupt the other coordinate trackers. Some radars can recognize that they have been deceived, and they then return to the remembered target trajectory with the intent of reacquiring the target echo. Therefore, a typical RGPO routine may consist of repetitions of the following sequence:

1. The pull-off phase;
2. Shutdown phase (leaving the tracker without a signal to track);
3. Repeat of pull-off phase to forestall reacquisition of the target.

A transponder or repeater jammer produces the false-target pulse for RGPO. The transponder is the same device discussed in Chapter 1 in connection with radar beacon operations, but is now working against an unfriendly radar. The repeater jammer is basically an amplifier that retransmits the incident radar pulse. Figure 3.8 indicates the minimum essentials of transponders and repeaters. The transponder shown uses a common receiving-transmitting antenna with a circulator as a duplexer. The detected radar pulse acts as a trigger to initiate the transmitted pulse.

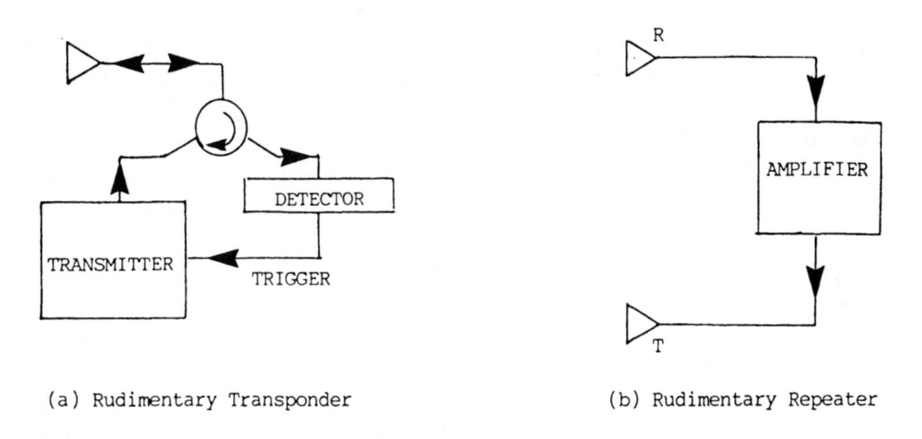

Figure 3.8 Transponder and repeater jammers.

Clearly, several additional features are required to make the jamming effective. Unless the jammer is to operate against a single-frequency, fixed-tuned radar, it must have a means of measuring the radar frequency and setting its transmitter to match that frequency. Manual tuning by an *electronic warfare officer* (EWO) has been practiced in the past, but the trend is toward full automation. Manual jammer adjustment is impractical in the case of jammers for single-seat aircraft and impossible for missile-borne jammers. Moreover, against a frequency-agile radar, manual tuning is impossible. The transponder of Figure 3.8(a) also lacks the means for selecting the radar pulse train (possibly from many received pulse trains) to respond to. This raises the topic of ESM capabilities touched on in Chapter 1. The jammer may be preprogrammed to respond to only the highest-priority threats, and, given those threat parameters, it may have a built-in or associated ESM capability enabling it to sort from all the received signals those that justify a response.

For the transponder, an increasing delay of the trigger, relative to the received pulses, clearly can affect the RGPO procedure. Figure 3.8(b) is not clear as to how such a procedure could be implemented for the repeater. The received pulse may be quite short, and, by the end of an RGPO cycle, we may want the jamming pulse to have been delayed by many pulsewidths relative to the skin-return pulse. The virtue of the repeater is its ability to retransmit a signal at the exact frequency of the radar with a phase bearing a consistent relation to the received phase. Thus, the frequency and phase of the incident radar pulse must be somehow remembered for use in forming the delayed jamming pulse.

Figure 3.9 shows one way of performing this function. We insert a short sample of the beginning of the incident pulse into a recirculating loop with delay equal to the length of the sample. At the loop output the recirculating samples appear head-to-tail to provide the carrier of the jammer pulse, which is initiated after the programmed delay. The memory loop contents are dumped after the jammer pulse transmission to prepare the loop for the next radar pulse. This scheme does produce a jamming pulse with the correct frequency and a consistent phase, both of which are essential if the jammer is to enjoy coherent interpulse processing gain (integration gain) against a coherent radar. The scheme has the fault that head-to-tail alignment of the recirculating samples does not, in general, produce the equivalent of a long continuous sinusoid at the carrier frequency. The loop delay is not, in general, an integral number of RF periods, so the loop output exhibits phase steps of magnitude:

$$\Delta\phi = \omega_0 \tau_d \text{ modulo } 2\pi$$

where ω_0 is the carrier frequency and τ_d is the loop delay; that is, $\Delta\phi$ is the departure of the loop phase shift from an integral multiple of 2π rad.

The transmitted pulse may encompass many recirculating samples. The stepped phase modulation can place much of the pulse energy at the edges of, or outside of, the victim radar receiver passband, thereby reducing the effectiveness of the jamming. Other memory techniques are coming into use, based on high-speed sampling of the incident waveform and storage for delayed playback, called *digital RF memory devices* (DRFMs). DRFMs avoid the head-to-tail addition of short RF samples, but spurious energy is still transmitted due to quantizing error and *analog-to-digital* (A/D) converter speed limitations.

The DRFM process typically involves the following steps:

1. Tuning a *local oscillator* (LO) to the approximate frequency of the intercepted signal;

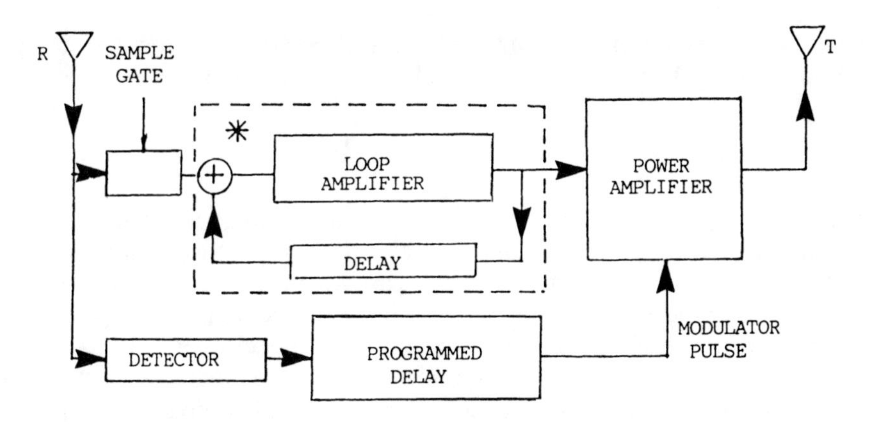

*These components may be replaced by DRFM in modern designs.

Figure 3.9 Repeater with frequency memory loop.

2. Down-converting the intercepted waveform to baseband, producing in-phase and quadrature (I and Q) components of the signal waveform (the LO of step 1 is used for the down-conversion);
3. Sampling and digitizing (A/D conversion) the I and Q signal components;
4. Digitally storing the I and Q waveforms;
5. Reconstituting the analog I and Q video by *digital-to-analog* (D/A) conversion at playback time;
6. Reconstituting an RF version of the original signal by means of single-sideband modulation, in which the I and Q video signals are the modulating waveforms. The LO of step 1 supplies the carrier for this up-conversion process.

For faithful reproduction of a signal of bandwidth B (Hz) and duration T (s), at least $2BT$ samples are required (BT in-phase samples and BT quadrature samples) [2, Chapter 3]. Modern radars employ coded waveforms with signal bandwidths from tens of megahertz up to several hundred megahertz, requiring sampling rates of comparable magnitude for the DRFM technique. The accuracy of sampling degrades as the sampling rate goes up. Moreover, the coarseness of quantizing grows as the time available for each A/D conversion decreases.[4] Typically, with current technology, a signal with B = 500 MHz would be sampled at a 1000 = MHz (1.0-GHz) rate and digitized with only 1- or 2-bit quantization, whereas a

[4]A/D conversion must keep pace with the sampling in order that the digitized version of the waveform be available for playback with a minimum of delay.

signal with B-50 MHz might be sampled at a 100 = MHz rate and digitized with 8-bit quantization. The minimum playback delay is likely to be 60 ns or more.

3.7 TARGET-JAMMER PULSE INTERACTION

At the beginning of the RGPO routine, the jammer delay is reduced to its minimum value so that the jamming pulse will fall within the victim radar's range gate. The minimum achievable delay will be at least several tens of nanoseconds, and perhaps 100 ns or more. In the region in which the jammer pulse overlaps the target return pulse, the two pulses add as phasors. If the jammer pulse is much larger than the target pulse, their relative phase is immaterial because the tracker will be dominated by the jammer. However, if the two pulses are of comparable magnitude, the relative phase can affect the effectiveness of the pull-off tactic. Figure 3.10 illus-

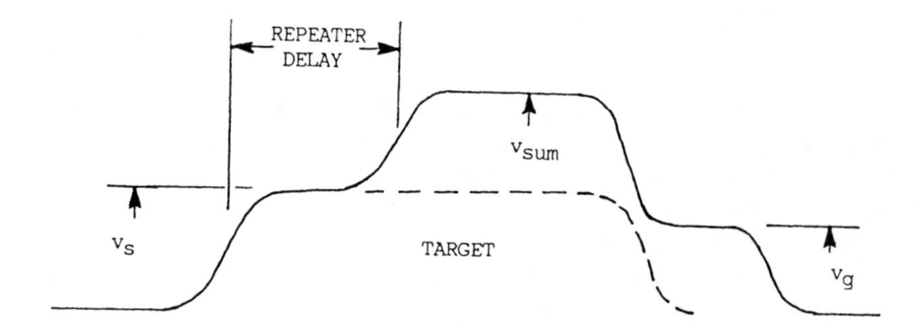

(a) Detected Target Plus Repeater Pulses

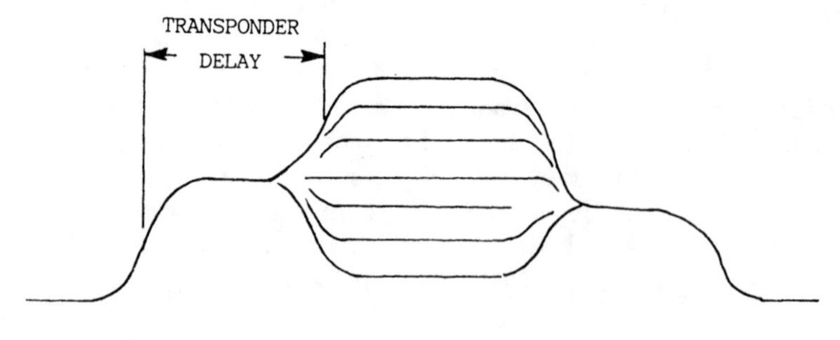

(b) Detected Target Plus Transponder Pulses

Figure 3.10 Target-jammer pulse interaction.

trates a case in which the jammer pulse is slightly below the level of the target return. For the repeater (Figure 3.10(a)) the relative phase is consistent from pulse to pulse, so the phasor addition in the overlap region repeats from pulse to pulse. In this picture the relative phase is not far from zero, for the envelope of the sum is not much less than the sum of the two envelopes. Had the phase been near 180°, the level in the overlap region would have been quite low.

In Figure 3.10(b), the relative phase of the transponder is random from pulse to pulse, so the display of several consecutive range traces reveals a flutter in the overlap region, with the envelope level ranging from a high equal to the sum of the two envelopes to a low equal to their difference. A typical tracking servo loop averages over many pulses, so we can estimate the effect of this random phase by averaging the phasor sum, assuming a uniform distribution of relative phase. The result is plotted in Figure 3.11. For J/S = 0 dB (equal pulse amplitudes), the average level of the sum, $\overline{V}_{sum}$, is about 2.2 dB higher than the signal amplitude V_S or the jammer amplitude V_j. When J/S reaches about 6 dB, the flutter (Figure 3.10(b)) produces an average that is only 0.7 dB greater than the jammer amplitude itself. When J/S drops to -6 dB, $\overline{V}_{sum}$ is only 0.7 dB above the level of the target return. Thus, beyond these ±6-dB limits, the split-gate error output is essentially under control of the dominant signal. Figure 3.11 demonstrates this fact.

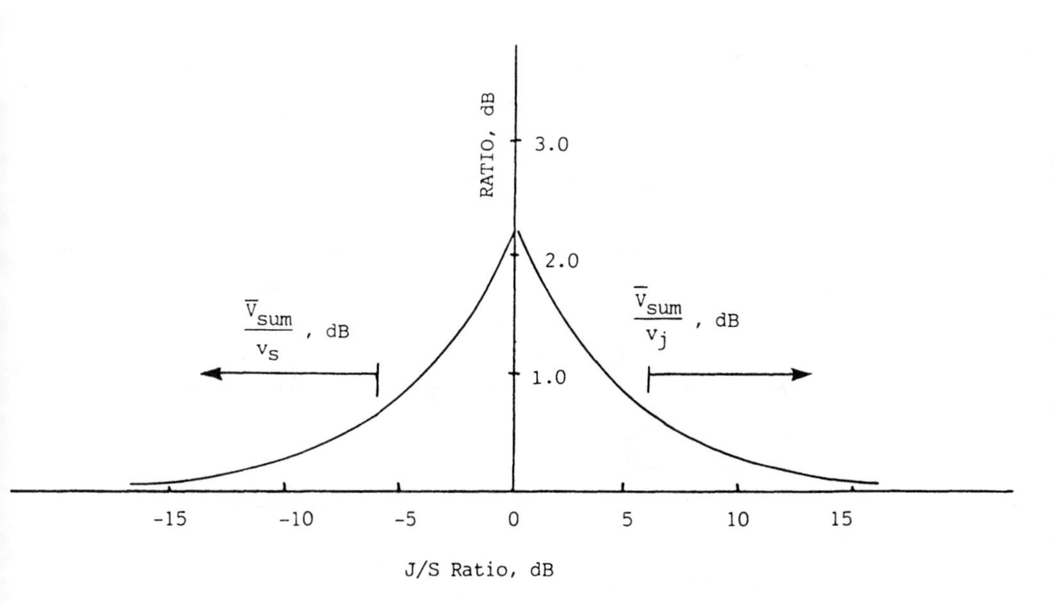

Figure 3.11 Average envelope amplitude, $\overline{V}_{sum}$, relative to dominant component amplitude (random phase).

3.8 CALCULATED *J/S* REQUIREMENTS

The item of greatest interest for RGPO is the motion of the range discriminator (error detector) null[5] as a function of jammer pulse pull-off distance. For very small *J/S* one expects the pull-off to move the null slightly in the direction of the jammer pulse, though never far enough to cause break lock; once the jammer pulse travels beyond the split gate, the gate snaps back to center itself on the target pulse. For higher values of *J/S*, one expects the null to lag behind the jammer pulse and finally to snap onto a lock on the jammer pulse once the target pulse has been left behind.

The motion of the null during the pull-off maneuver is easy to compute, irrespective of tracking loop considerations. One can then conclude that, provided the pull-off is executed slowly enough, the tracking servo will follow the motion of the null. Whether the tracker will hold to the null under faster pull-off maneuvers will depend on the tracking servo transient response.

We carried out computations for the idealized pulse shapes of Figure 3.12. The rise and fall times of the trapezoidal signal pulse were each 10% of the nominal

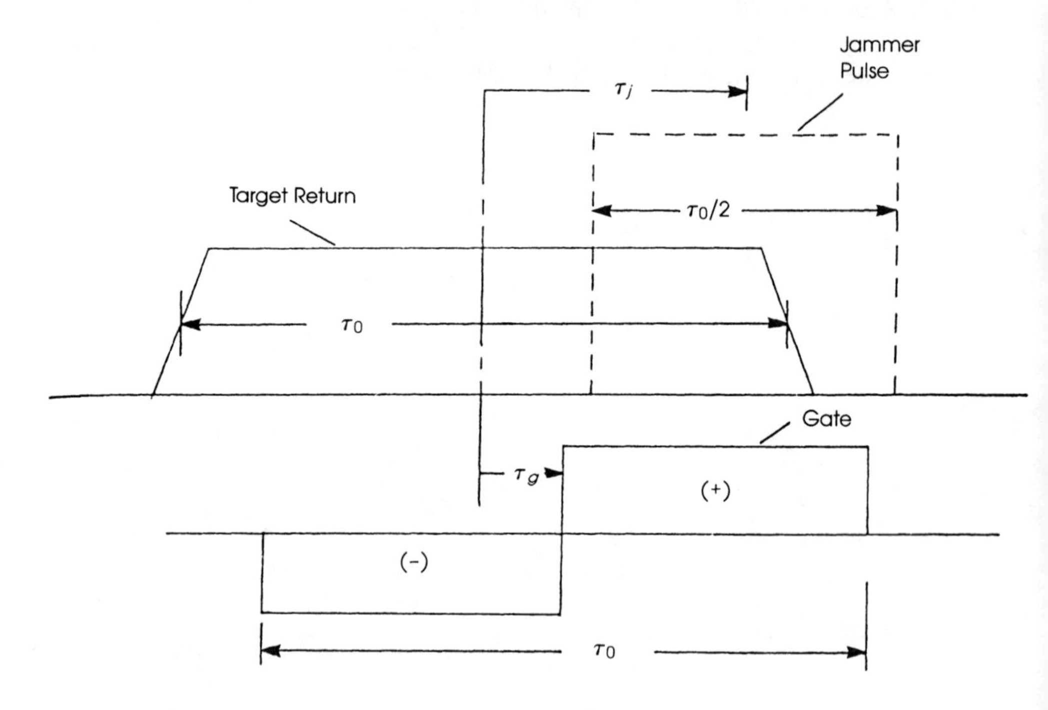

Figure 3.12 Jammer pulse target in range gate.

[5]Assuming that the split gate follows the null.

pulse duration τ_0. The split gate of the discriminator had a width τ_0. The jammer pulse had a width of $\tau_0/2$. We assumed that the jammer was a repeater, resulting in a fixed relative phase between jammer and signal pulses. The motion of the null as a function of jammer pull-off distance is plotted in Figures 3.13, 3.14, and 3.15 for relative phase values of 0°, 90°, and 180°, respectively. For zero phase difference, we see (Figure 3.13) that, even for J/S = 0 dB, the jammer succeeds in carrying the null with it, although with a considerable lag until the gate pulls free of the target. As J/S increases, the lag becomes less, until at infinite J/S the null follows precisely (the dashed line) the motion of the jammer pulse. With 90° relative phase a J/S of 0 dB is insufficient to effect break lock. The null is pulled to almost half a pulsewidth, but at that point the target's hold on the gate is more than a match for the jammer. The jammer continues to pull away, and the gate snaps back to center itself on the target. For J/S = 3 dB and higher, the pull-off does succeed. For 180° relative phase, at J/S = 6 dB, as soon as the jammer pulse moves slightly to the right (Figure 3.12), the null moves to the left of the center of the signal pulse because of the partial cancellation in the overlap region. The null returns to the plot (that is, it reaches the point where it is centered on the target pulse) only after the jammer pulse has moved 0.2 pulsewidth. In this case the pull-off succeeds for all $J/S \geq 6$ dB.

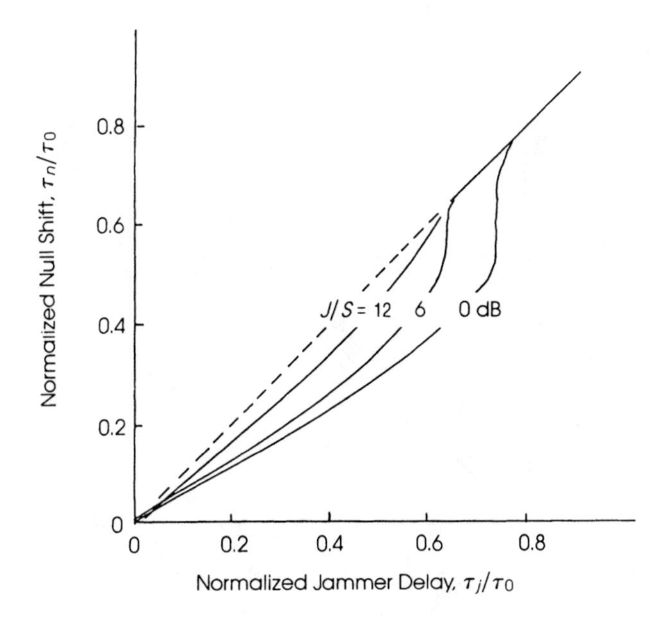

Figure 3.13 Range gate pull-off (relative phase = 0°).

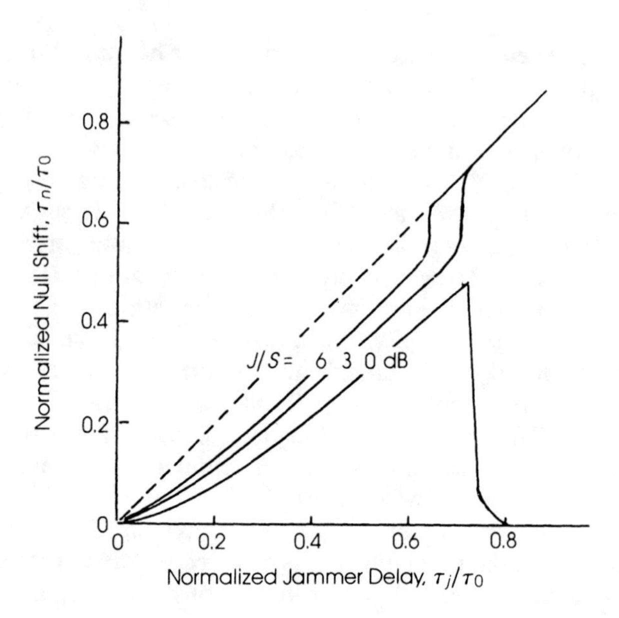

Figure 3.14 Range gate pull-off (relative phase = 90°).

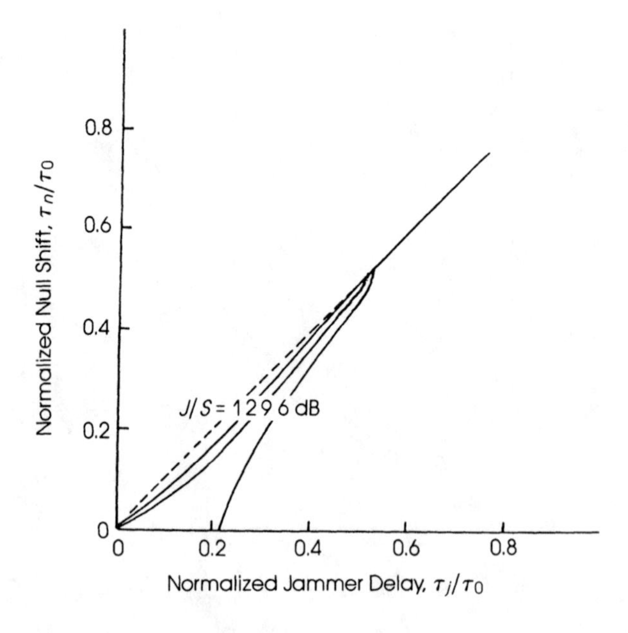

Figure 3.15 Range gate pull-off (relative phase = 180°).

3.9 SIMULATION RESULTS

A number of RGPO simulations were run to illustrate the impact of J/S and pull-off rate on the success of this form of ECM. In the model, both the signal and jammer pulses are rectangular with duration τ. The range tracker employs a split gate of total duration 2τ. The range servo is type 1 with undamped natural frequency $\omega_n = 2\pi$ rad/s and a damping ratio $\zeta = 0.5$ (see Appendix A). In the simulation, time displacements of the target pulse, the jammer pulse, and the range gate are normalized to the pulsewidth τ. Thus, in the plots, the vertical axes are scaled to give jammer delay[6] t_j and gate position t_g in units of τ.

In Figures 3.16 through 3.18, the jammer pull-off is at a rate of 1.33τ per second.[7] To relate this rate to the range rate $\dot{R}$ of the simulated false target, we suppose that the radar employs a simple pulse of duration $\tau = 0.5\ \mu s$. This corresponds to a range interval of

$$\Delta R = \frac{c\tau}{2} = 75 \text{ m}$$

so the simulated range rate is

$$\dot{R} = 1.33 \times 75 = 100 \text{ m/s}$$

or about 194 knots. With a J/S of 0 dB (Figure 3.16), the RGPO maneuver fails. The range gate is nudged about half a pulsewidth, and then it falls back to its lock on the target. At $J/S = 3$ dB (Figure 3.17), the jammer succeeds in carrying away the gate. The transient lag $(t_j - t_g)$ reaches a maximum value[8] of about 0.7τ. The steady-state lag (measured after about 4 s) is

$$(t_j - t_g)_{ss} = \frac{\text{pull-off rate}}{k_v} = 0.16\tau$$

In this equation, the static error coefficient, k_v (see Appendix A), is made equal to 2π when the pulse being tracked lies entirely inside the range gate. As the pulse begins to pull away from the range gate, the AGC in the range-tracking loop raises the loop gain, thus increasing the value of k_v. At $J/S = 10$ dB (Figure 3.18) the transient lag is smaller, and the steady state is reached in about 2.5 s.

---

[6]In the simulation, target pulse position, t_s, was held at zero (stationary target), so t_j and t_g in the plots give jammer and gate positions relative to the position of the target return.

[7]The tracker is initially locked to the target return. RGPO begins at $t = 1.0$ s. The slope of the t_g *versus* t plot is equal to the pull-off rate. Note that not all vertical scales are identical in the various plots.

[8]Should the lag reach 1.5τ, the jammer pulse would be lost to the gate.

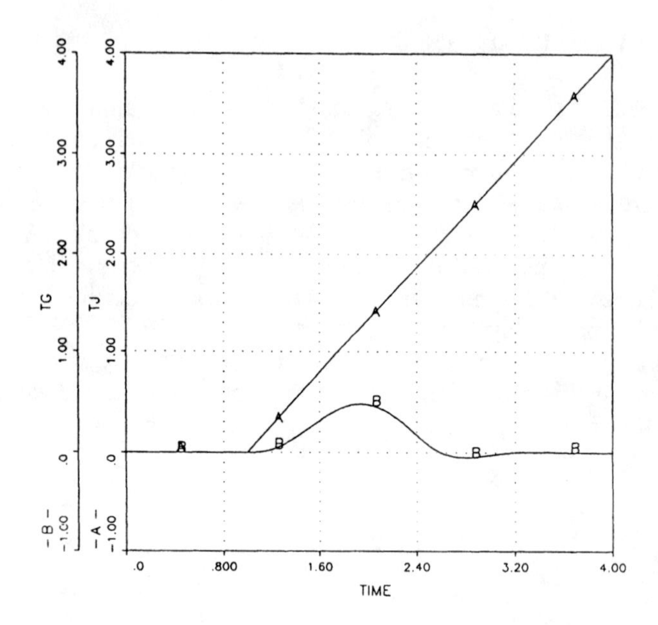

Figure 3.16 RGPO attempt; rate = 1.33τ/s, J/S = 0 dB.

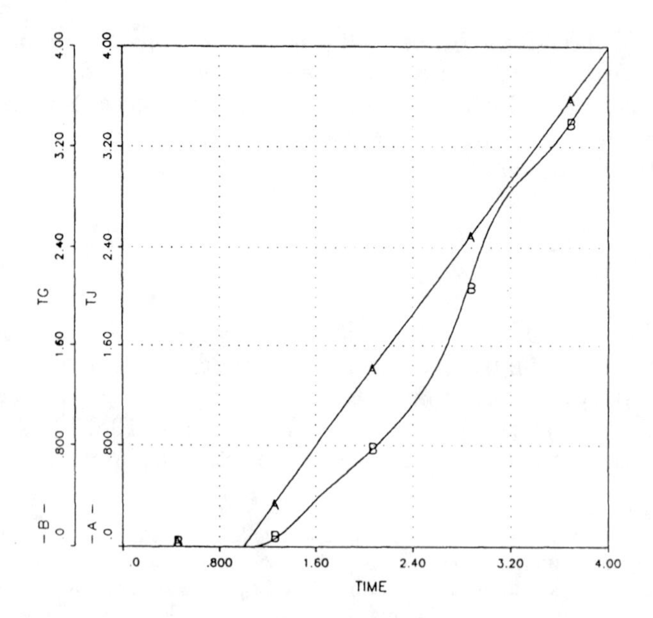

Figure 3.17 RGPO attempt; rate = 1.33τ/s, J/S = 3 dB.

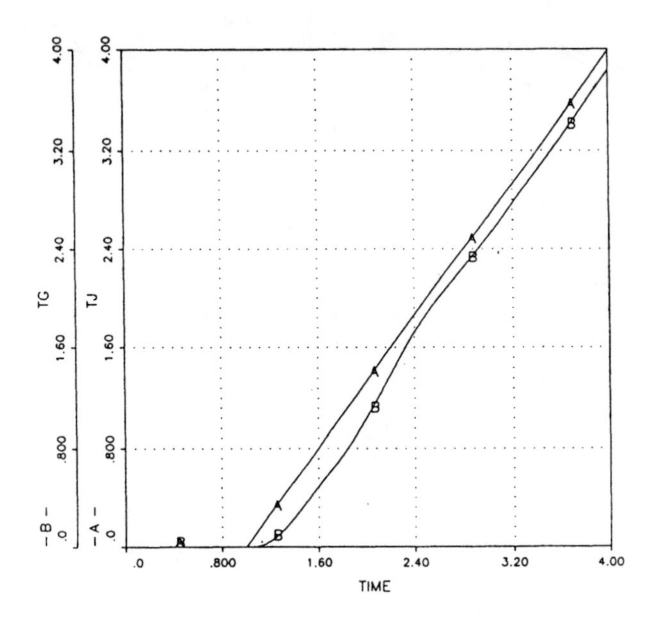

Figure 3.18 RGPO attempt; rate = 1.33τ/s, J/S = 10 dB.

In Figures 3.19 through 3.21 the pull-off rate is doubled (2.67τ/s). In terms of the previous example, this represents a range rate of $\dot{R} = 200$ m/s, or about 388 knots. Now even with $J/S = 6$ dB (Figure 3.19), the RGPO maneuver fails; the gate is moved about 0.75τ, but it falls back to a lock on the target. At $J/S = 10$ dB (Figure 3.20), the jammer succeeds in dislodging the gate from the target, but fails to carry it away. The gate comes to rest at $t_g = 2.7\tau$. The RGPO maneuver might be considered only marginally successful, for the tracker could reacquire the target by executing a narrow search extending only $\pm 2\tau$ around its resting point. At J/S = 15 dB (Figure 3.21), the RGPO succeeds, with a steady-state lag of

$$(t_j - t_g)_{ss} = \frac{\text{pull-off rate}}{k_v} = 0.27\tau$$

being achieved after about 2 s.[9]

[9]As stated earlier, k_v is not a constant in this model, because AGC action begins raising the track loop gain as soon as the pulse moves off range-gate center.

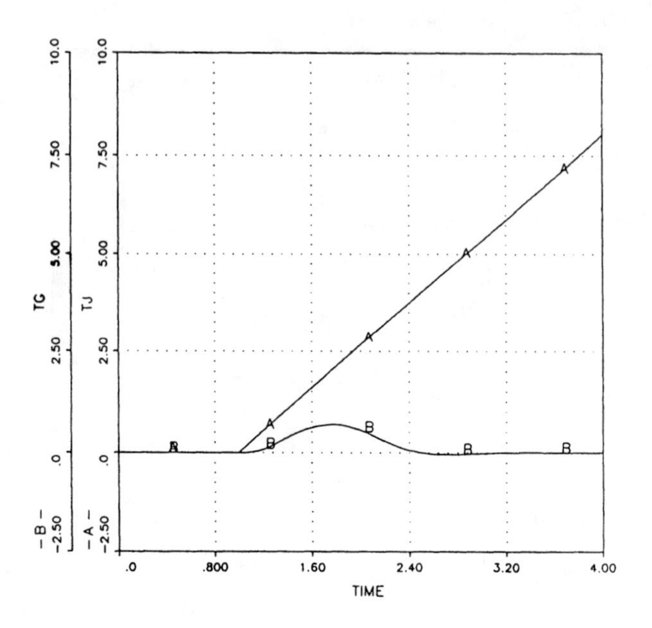

Figure 3.19 RGPO attempt; rate = 2.67τ/s, J/S = 6 dB.

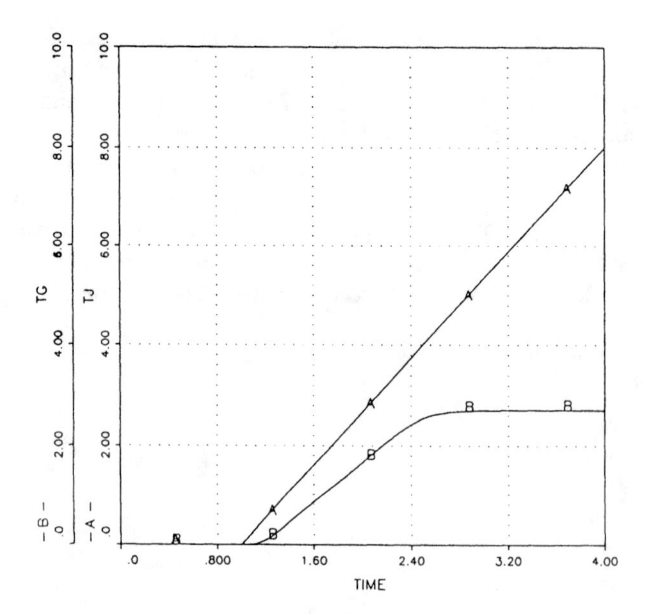

Figure 3.20 RGPO attempt; rate = 2.67τ/s, J/S = 10 dB.

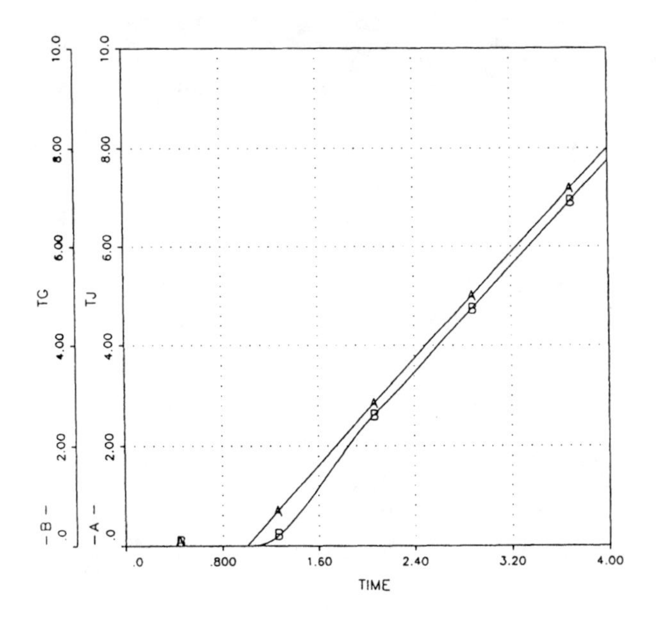

Figure 3.21 RGPO attempt; rate = 2.67τ/s, J/S = 15 dB.

3.10 LEADING-EDGE TRACKING

Figure 3.22 illustrates the three distinct portions of a combined signal plus jammer pulse when the jammer delay is a minimum:

1. The early portion consisting of signal alone;
2. The overlap region consisting of signal plus jamming;[10]
3. The final portion consisting of jamming alone.

If a tracker can be devised to accept only the early segment (1), it is possible to nullify the RGPO tactic. A narrow split gate alone is insufficient. If such a gate were initially set on the very leading edge (Figure 3.22(a)) it would be driven to the right, seeking an area balance. Then, on reaching the second rise (the edge of the jammer pulse), it would seek an equilibrium point still further to the right and come to rest (or drift about) in the wide, flat, overlap region. It would then be carried away by the jammer pull-off. If the waveform is differentiated, the result may resemble Figure 3.22(b), and a narrow split gate may then be locked to the leading-edge derivative pulse, as indicated with the jammer pulse completely ignored.

[10]If, as in Figure 3.10(b), there is a random variation of the pulse-to-pulse relative phase, this overlap region exhibits an amplitude flutter.

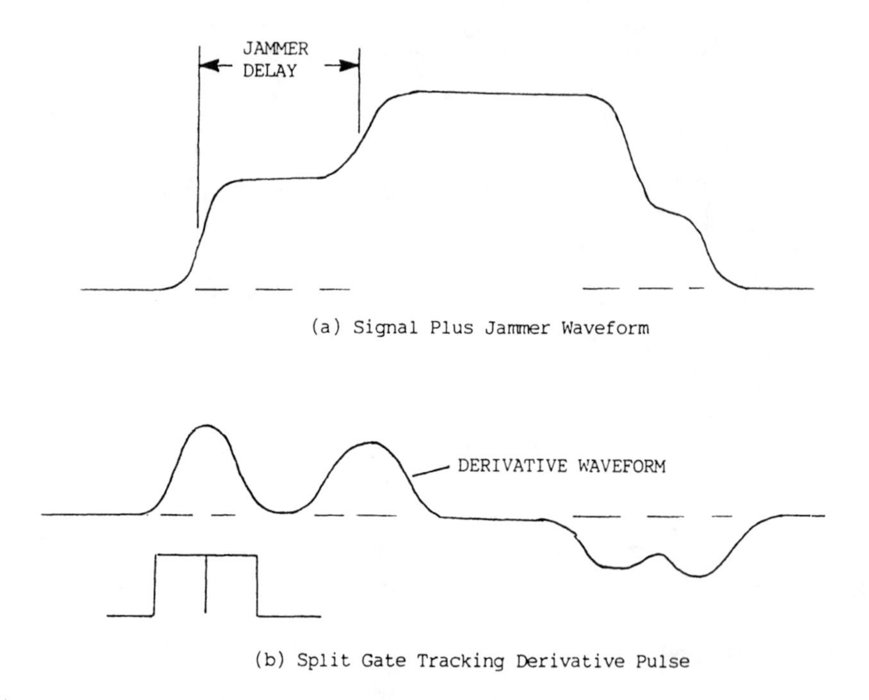

Figure 3.22 Leading-edge tracking on differential video.

It is clear in Figure 3.22 that the jammer will be able to defeat the leading-edge tracker only if it can reduce its delay sufficiently to place enough of the jammer leading-edge pulse in the split gate to overcome the influence of the target. To illustrate this situation, we constructed a simple model in which the leading edges of both the signal and jammer pulses were half-cosine waveforms; that is,

$$v(t) = \frac{A}{2}\left(1 - \cos\left(\pi \frac{t}{\tau_R}\right)\right) \qquad \text{for } 0 \leq t \leq \tau_R$$

where A is the amplitude of the flap top of the main body of the pulse, and τ_R is the rise time (from zero to flat region). The resultant leading-edge pulses, the derivatives of the leading edges, are of the form:

$$v_{\mathrm{LE}}(t) = \frac{\mathrm{d}v}{\mathrm{d}t} = \frac{\pi A}{2\tau_R} \sin\left(\pi \frac{t}{\tau_R}\right) \qquad \text{for } 0 \leq t \leq \tau_R$$

This results in the situation described by Figure 3.23(a), in which the jammer leading-edge pulse is shown with a delay τ_j relative to the signal leading-edge pulse. Both

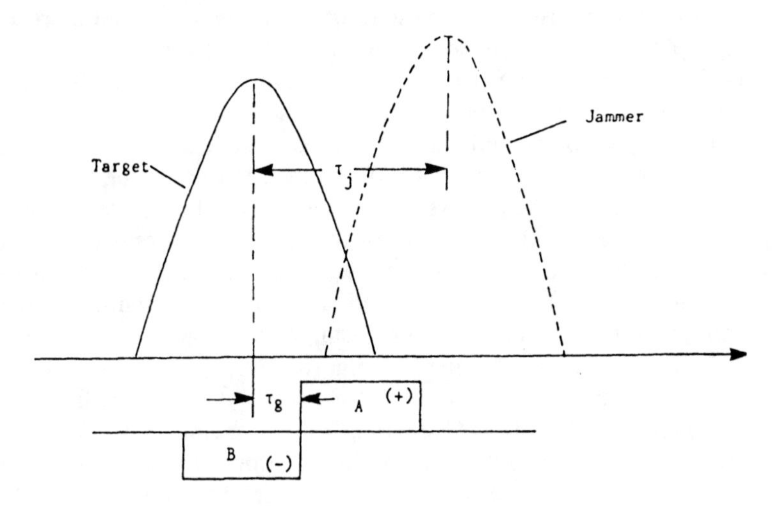

(a) Leading Edge Pulses and Gate

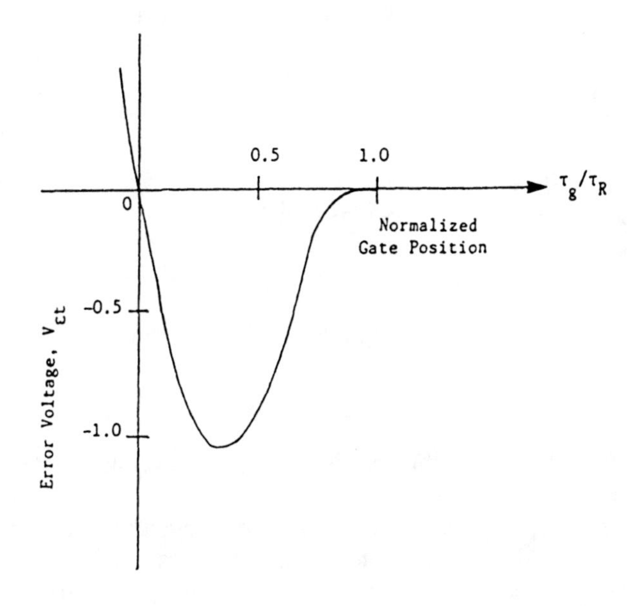

(b) Target Error Voltage as a Function of Gate Position

Figure 3.23 Leading-edge tracking.

pulsewidths are equal to τ_R, the original pulse rise time. The leading-edge split gate of width τ_R is shown at a position τ_g to the right of the signal pulse. Figure 3.23(b) is a plot of the split-gate error detector output, v_ϵ, as a function of τ_g/τ_R when only the signal is present. It reaches a peak negative value at $\tau/\tau_R = 0.352$. Our objective is to determine the value of J/S that is required for RGPO.

As would be expected, we find that the J/S required depends on the minimum achievable jammer delay $\tau_{j\min}$. The greater the delay, the smaller the portion of the jammer pulse falling within the gate when the gate is locked to the target and, consequently, the larger must be the J/S to dislodge the gate. To simplify the problem, we shall assume the signal and jammer carriers to be in phase. Now imagine the following experiment. We make the signal and jammer pulse amplitudes equal and fix τ_j. Beginning with the gate centered on the target, we vary τ_g and plot the error voltage $v_{\epsilon s} = v_{\epsilon s}(\tau_g)$ for the signal alone (see Figure 3.23(b)). We then carry out the same process for the jammer alone and plot $v_{\epsilon j} = v_{\epsilon j}(\tau_g)$. This plot has the same shape as the $v_{\epsilon s}$ plot, but it is shifted to the right by τ_j. For small τ_j we find that over a range of values of τ_g, the two error signals are of opposite polarity; $v_{\epsilon s}$ tends to restore the gate to its position on the target, and $v_{\epsilon j}$ tends to pull it away (toward the jammer pulse).

For each value of τ_g, we can compute the ratio:

$$\rho_\epsilon = \left| \frac{v_{\epsilon s}}{v_{\epsilon j}} \right|$$

For any given gate position, the track loop can be in equilibrium only if J/S is raised from unity to a value that makes

$$|v_{\epsilon s}| = |v_{\epsilon j}|$$

Thus, the required J/S for equilibrium is

$$J/S = 20 \log \rho_\epsilon$$

Figure 3.24 is a plot of this required J/S as a function of gate position for a number of values of jammer delay. In all cases, the required J/S becomes infinite as τ_g approaches τ_j, for that condition calls for the gate to be centered on the jammer, which, as noted earlier, cannot be achieved as long as some of the signal pulse remains in the gate. We note that for $\tau_j \geq 0.8\tau_R$ the curves exhibit humps at low values of τ_g and then fall away before rising again near $\tau_g = \tau_j$. To achieve RGPO, the jammer must produce a J/S high enough to get over the hump. Once over the hump, the J/S is excessive for equilibrium, and the tracker will shift the gate to the right until equilibrium occurs on the rising portion of the curve. Then the jammer can increase its delay and carry away the gate.

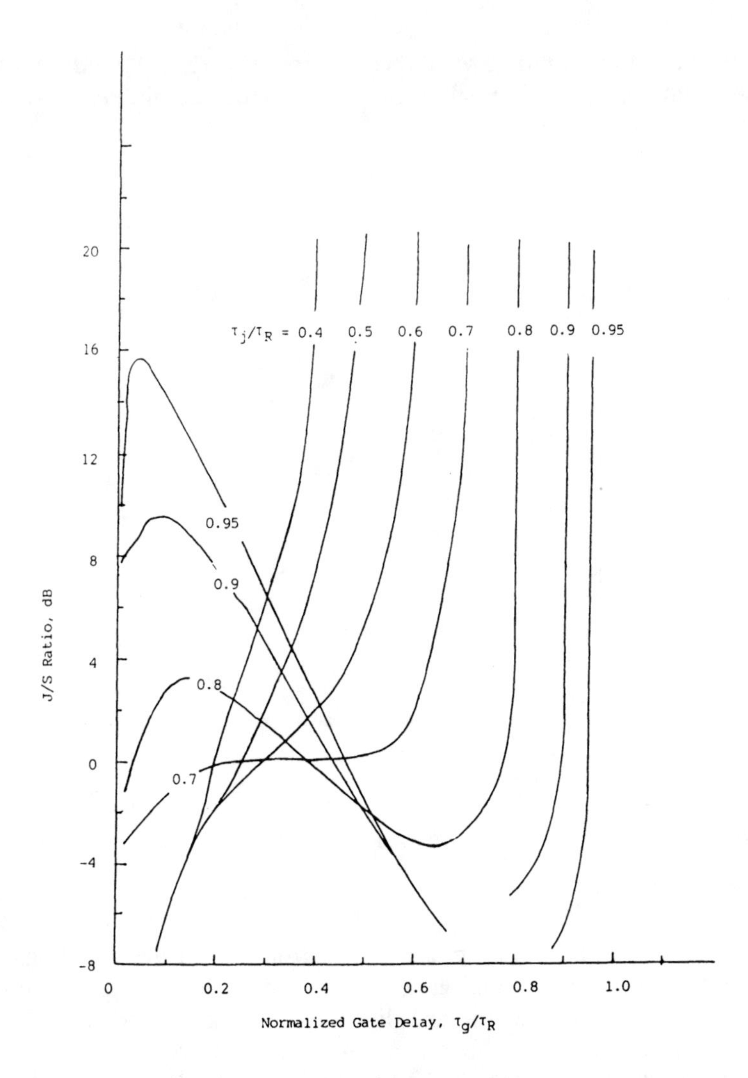

Figure 3.24 *J/S* required for equilibrium at gate position τ_g for various values of jammer delay τ_j.

Figure 3.25 is a plot of the *J/S* required for RGPO as a function of jammer delay. For small values of τ_j/T_R, a *J/S* of slightly over 0 dB suffices, but as τ_j approaches τ_R the required *J/S* becomes infinite, as expected. Clearly, with leading-edge tracking on fast rise-time pulses, a repeater or transponder jammer will find it difficult to achieve RGPO. The somewhat unrealistic assumption concerning the phase of the jammer pulse relative to the signal pulse brings into question the portion of Figure 3.25 for low values of τ_j, but at values of τ_j that demand a *J/S* of several dB the curve should be valid, because (see Figure 3.11) under these conditions the signal plus jammer envelope amplitude is then dominated by the jammer, and relative phase is unimportant.

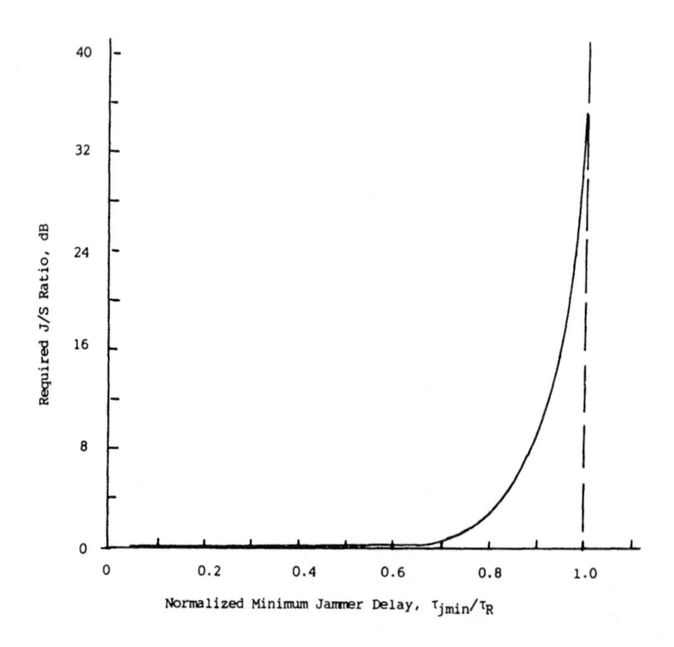

Figure 3.25 Required *J/S* for RGPO as a function of jammer delay. (leading-edge tracker model)

The presence of a jammer pulse much stronger than the signal has an additional potential effect. If the range gate associated with the AGC is of a width equal to the expected signal pulsewidth (as is likely to be the case), the powerful jammer pulse will dominate the AGC, perhaps forcing the receiver gain so low that the weaker signal pulse is too weak for leading-edge tracking. We can avoid this situation by employing an AGC gate of width comparable to the width of the leading-edge gate whenever leading-edge tracking is in progress. The range-gate width for other receiver channels might or might not be shrunk under these circumstances. For instance, if angle tracking is based on monopulse techniques, there is no harm in allowing the jammer pulse to come through a full (normal)-width range gate into the angle channels, for the jammer serves as a perfectly satisfactory source for angle tracking. This would not be true of other angle-tracking techniques, such as conical scan. If the jammer were to initiate AM countermeasures aimed at disrupting a *conical scanning* (CONSCAN) tracker before RGPO begins,[11] it would be advisable to shrink the range gate associated with the angle tracker in order to exclude the jammer from the angle channel.

Manual leading-edge tracking can defeat the RGPO tactic. The operator simply restrains the tracker from following the jammer pull-off. The operator can also override the AGC and keep the target pulse at a visible level. Unfortunately, in many systems the human operator is absent or cannot afford to give full attention to a single threat, for the system may have multiple targets in track.

3.11 RGPO AGAINST FM-CW RADAR

Although most radars determine range by directly measuring the time delay of an echo pulse, certain radars operate in a CW mode. A constant-frequency sinusoid offers no potential for ranging, but it is an excellent waveform for measuring target doppler. When range measurement is a necessity, the carrier is frequency-modulated in a periodic fashion. Common frequency *versus* time patterns are audio-frequency sinusoidal, triangular, or sawtooth waveforms. Target range is extracted as a beat frequency between the transmitted waveform and the echo [1, Chapter 16]. The beat, or difference, frequency occurs because at any instant the CW echo, due to its round-trip time delay, is different in frequency from the frequency being transmitted. For a distant target to execute RGPO against such a radar, the jammer must be able to imitate a CW echo coming initially from the target range, then from progressively increasing or decreasing ranges (because the CW waveform is repetitive, false targets at ranges less than the target range can be generated). Van Brunt [3] has suggested a method for producing the required false-target echo. The jammer locks an oscillator to the mean frequency of the radar (the carrier frequency). At the same time, it employs a frequency discriminator to extract the frequency *versus* time pattern of the radar's modulation. By a serrodyne [3] process, synchronized to the demodulated FM waveform, the jammer produces false-target FM on the reconstituted carrier, initially mimicking a target at the jammer range. Gradually, the serrodyne pattern is modified to mimic a target at slowly changing range.

Because a CW radar transmits and receives simultaneously, it faces the very difficult problem of achieving sufficient isolation between its transmitting and receiving antennas. Weak target returns must be detected in the presence of leakage of transmitted energy into the receiver. Because of the round-trip path loss, the dynamic range:

---

[11] If the jammer does combine AM against the angle tracker with RGPO, the AM is likely to be delayed until the jammer has pulled clear of the skin return.

$$\rho_R = \frac{\text{radar ERP}}{\text{received echo power}}$$

is enormous. The jammer also faces the problem of simultaneous reception and transmission, but because it receives (intercepts) the radar emission over a one-way path, its corresponding dynamic range ratio:

$$\rho_J = \frac{\text{jammer ERP}}{\text{intercepted radar signal power}}$$

is very much less. The ratio of the two dynamic range factors is

$$\frac{\rho_R}{\rho_J} = \left(\frac{4\pi R^2}{\sigma_T}\right)^2 \frac{1}{J/S}\left(\frac{A_J}{A_R}\right)$$

where R is the range separation, σ_T is the target *radar cross section* (RCS), A_J and A_R are the effective capture areas of the jammer and radar receiving antennas, and J/S is the jammer-to-signal power ratio that the jammer wishes to achieve against the radar. If the target RCS is a square meter, the first factor contributes 180 dB to the ρ_R/ρ_J ratio at $R = 10$ km. The reciprocal of the required J/S will probably only contribute -10 to -20 dB or less, and A_J/A_R is unlikely to be lower than -30 dB, so it is clear that the jammer's isolation problem is mild compared with that of the radar.

REFERENCES

1. Skolnik, M.I., ed., *Radar Handbook,* 1st Ed., McGraw-Hill, New York, 1970.
2. Schwartz, M., *Information Transmission, Modulation, and Noise,* McGraw-Hill, New York, 1959.
3. Van Brunt, L.B., *Applied ECM,* EW Engineering, Dunn Loring, VA, 1978.

Chapter 4
ANGLE DECEPTION

4.1 BACKGROUND

The lethality of guided weapons and ballistic weapons such as *antiaircraft artillery* (AAA) depends strongly on the ability of the fire control system to obtain accurate measurements of the target's angle coordinates. Countermeasures against angle measurement and tracking are therefore crucial to the survival of military air and surface vehicles that may be targeted by radar-based fire control systems.

Fire control radar systems rely on automatic trackers for continuous measurement of target angle. The angle measurement, as is true of the other coordinate measurements (Appendix B), depends on the existence of a resolution function (the antenna beam pattern) that has peak response occurring at the angle of the target.

For tracking in one angle coordinate, two identical overlapping beams squinted on either side of the boresight axis yield equal responses when the boresight axis aims at the target. The difference of the responses goes to zero with the target on boresight, and it takes on opposite polarities for target positions on opposite sides of the boresight. When the two beams are present simultaneously, as in monopulse trackers, the tracking method is called *simultaneous lobing.* If the two positions are occupied alternately by a single beam, as when the beam of an array antenna is switched back and forth, the method is called *sequential lobing.*

Early angle tracking was performed manually, with an operator "dithering" the beam back and forth slightly to trace out the peak of the beam-response function, and keeping the dither centered on the peak. Conical scan is a two-dimensional automated form of dither in which a pencil beam with circular symmetry executes a scan about the tracker's boresight axis. The TWS mode combines target angle measurement with searching. The beam-response function for a target at a given angle is traced out only intermittently, once per scan, so the target tracker functions by maintaining a TWS time gate centered on the time position of the peak target response within the scan.

Sequential-lobing schemes (including both discrete lobe switching and continuous scanning) base their estimates of target position on a comparison of signal amplitudes received at slightly different times. They are therefore susceptible to measurement errors caused by both additive interference, with amplitude fluctuating with time, and scintillation of the amplitude of the target return. Angle countermeasures against this category of radars rely on amplitude-modulated jamming to induce angle errors. Deceptive angle countermeasures against these radars, aiming to mislead the tracker as to actual target angle, need to synchronize their modulation patterns with the motion of the radar beam. When the radar employs a single beam for both transmission and reception, the jammer is able to synchronize to the motion of the detected transmitted beam. An effective radar countermeasure against such jammers is to employ separate transmitting and receiving antennas, with the nonscanning transmitting antenna providing constant illumination at the target. Such systems are referred to as *lobe-on–receive-only* (LORO) or *scan-on–receive-only* (SORO) systems. Against LORO or SORO radars the AM jammer is forced to operate blindly, with far less effect than if synchronized.

Monopulse, the simultaneous lobing form of angle measurement, depends not on a time variation of received signal amplitude but on the relative amplitudes (or phases) of the target return received simultaneously in a pair of beams. The measurement is unaffected by time variation of the received target return amplitude. A jammer emission from the target vehicle provides an equally acceptable (and very likely stronger) signal for monopulse angle measurement. Amplitude-modulated jamming is therefore ineffective against monopulse. Section 4.6 discusses two schemes that have been devised for monopulse angle deception. Finally, Section 4.7 discusses conopulse or *scan-while-compensation* (SWC). This is a sort of hybrid combination of monopulse and conical scan.

4.2 ANGLE DECEPTION AGAINST CONICAL SCAN

Figure 4.1 represents a pattern of four overlapping antenna lobes centered on target T. These lobes might be produced by lobe switching of an array antenna, alternating between beam positions 2 and 4 to provide an azimuth measurement and between positions 1 and 3 to provide an elevation measurement. Figure 4.2 depicts the continuous rotation of a symmetrical pencil beam over a circular path about target T, including the beam positions that would be occupied in the lobe-switching scheme. This is the motion of a CONSCAN beam. The beam axis, squinted a fraction of a beamwith, θ_s, off the scan axis, rotates at a constant rate about the scan (boresight) axis and sweeps out a cone of half-angle θ_s. Rapid scanning (many scans per second) is desirable, both for obtaining frequent samples of target dynamics and for cancellation of target amplitude scintillation components in the frequency range below the scan frequency.

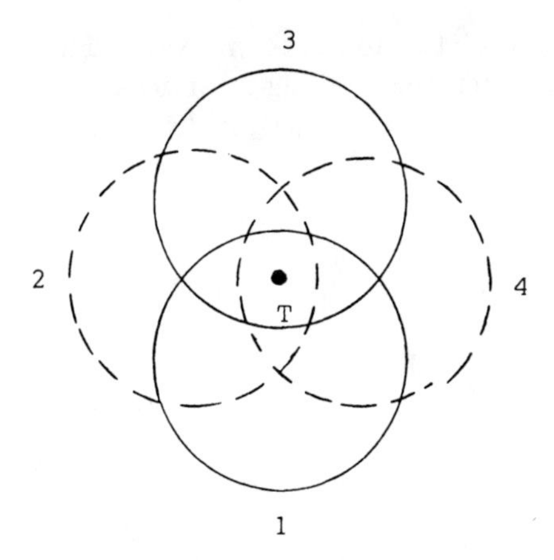

Figure 4.1 Discrete beam positions in lobe switching.

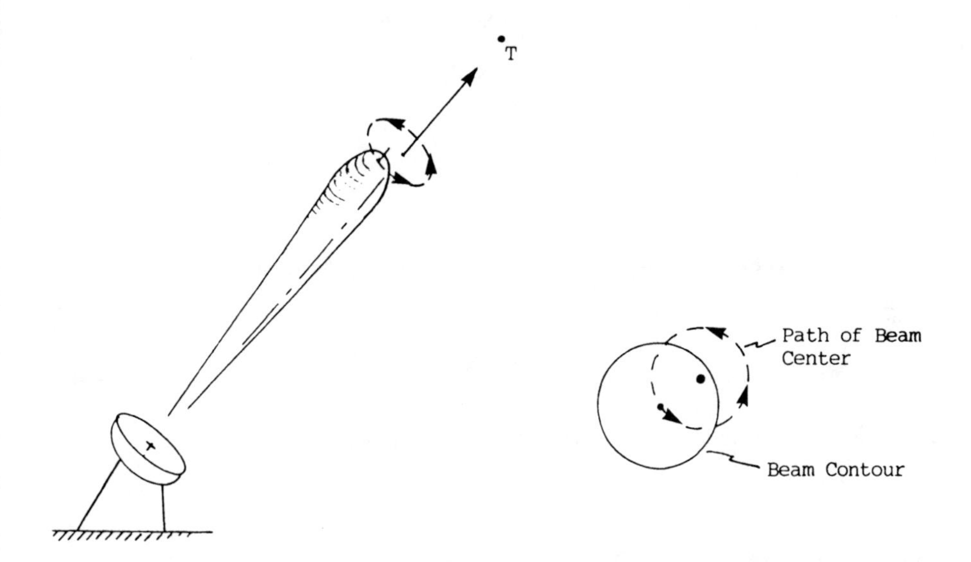

Figure 4.2 CONSCAN beam motion.

Figure 4.3 illustrates the CONSCAN beam motion in the plane normal to the scan axis, which contains target T (the same plane as Figure 4.1). The radar range to this plane is R. Point S is the spot where the scan axis pierces the plane. Point A is where the beam axis pierces the plane, and point T is the target location. The tracking error is θ_T, the departure of the scan (boresight) axis from the target, and

θ is the angle of the target relative to the beam axis. In Figure 4.3, this angle varies over the scan cycle; but when the tracking error goes to zero, θ remains constant, equal to squint angle θ_s. The axis α defines the direction of the tracking error.

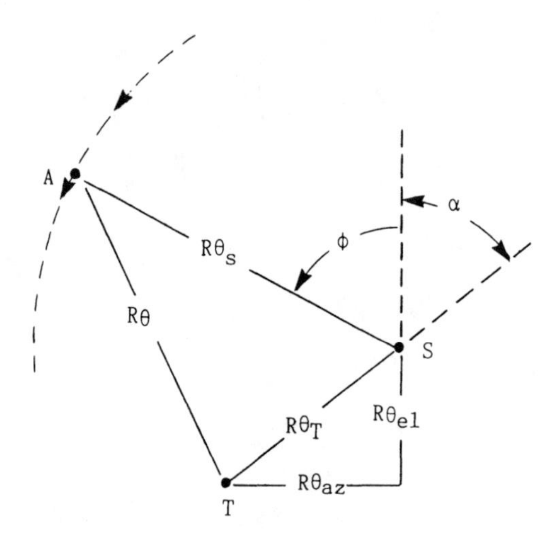

Figure 4.3 CONSCAN geometry.

Now, θ_T has azimuth and elevation components given by

$$\theta_{\text{az}} = \theta_T \sin(\alpha) \tag{4.1}$$

$$\theta_{\text{el}} = \theta_T \cos(\alpha) \tag{4.2}$$

Scan angle, ϕ, is measured from the vertical, and its value is

$$\phi = \omega_s t$$

where ω_s is the scan rate (rad/s). Because angles θ, θ_T, and θ_s are small, the sides of triangle SAT are given by $R\theta_s$, $R\theta$, and $R\theta_T$. The law of cosines applied to this triangle yields the following relation between the angles:

$$\theta^2 = \theta_T^2 + \theta_s^2 + 2\theta_T\theta_s \cos(\alpha + \phi) \tag{4.3}$$

As noted, the beam has circular symmetry, so its normalized one-way voltage pattern can be expressed as

$$g_l = g_l(\theta)$$

If the same beam is used for transmitting and receiving, the target return amplitude is

$$v_s = v_0 g_1^2(\theta) \tag{4.4}$$

where v_0 is the target return amplitude on the beam peak. For a specific beam pattern, $g_1(\theta)$, we can determine the scan modulation envelope of the target return, as a function of scan angle ϕ, by substituting the value of θ from (4.3) into (4.4). As might be expected, the modulation envelope is nearly sinusoidal for small θ_T, and the modulation disappears where the tracking error is reduced to zero.

To illustrate the scan modulation envelope, let us approximate the main-lobe one-way voltage pattern by

$$g_1(\theta) = e^{-a\theta^2} \tag{4.5}$$

where

$$a = 2\,\frac{\ln(2)}{\beta^2} \tag{4.6}$$

and β is the 3-dB beamwidth. The two-way pattern is then

$$g_1^2(\theta) = e^{-2a\theta^2}$$

With the value of θ^2 from (4.3) inserted in this equation, we have

$$g_1^2(\theta) = e^{-2a(\theta_T^2 + \theta_s^2)} e^{-4a\theta_T\theta_s \cos(\phi + \alpha)} \tag{4.7}$$

The first factor contains no scan modulation. For small θ_T, this first factor gives the mean value of the round-trip gain. When $\theta_T = 0$, this factor gives the *crossover loss*, $g_1^2(\theta_s)$, that results from the finite squint angle. The scan modulation is contained in the second factor (recall that $\phi = \omega_s t$). Let us then write the scan-modulated target return envelope as

$$v_s = v_0 e^{-4a\theta_T\theta_s \cos(\omega_s t + \alpha)}$$

where v_0 is the mean value of the envelope. For small tracking error θ_T, we can approximate the exponential function by the first two terms of its Taylor series:

$$v_s \approx v_0[1 - 4a\theta_T\theta_s \cos(\omega_s t + \alpha)]$$

We extract the azimuth and elevation error signals for the angle servos from v_s by synchronous detection, using the quadrature scan-frequency references $\sin(\omega_s t)$ and $\cos(\omega_s t)$, as indicated in Figure 4.4. Multiplying by $\sin(\omega_s t)$ gives

$$\begin{aligned} v_{\sin} &= v_s \sin(\omega_s t) \\ &= v_0[\sin(\omega_s t) - m \cos(\omega_s t + \alpha) \sin(\omega_s t)] \\ &= v_0\left\{\sin(\omega_s t) - \frac{m}{2}[\sin(2\omega_s t + \alpha) - \sin \alpha]\right\} \end{aligned} \tag{4.8}$$

where m represents the modulation index, $4a\theta_T\theta_s$. The low-pass filter (Figure 4.4) retains only the dc term, v_{az}, of the preceding equation:

$$v_{az} = v_0 \frac{m}{2} \sin(\alpha)$$

With the value of $\sin(\alpha)$ from (4.1) substituted in this expression, we have

$$\begin{aligned} v_{az} &= v_0 \frac{m\theta_{az}}{2\,\theta_T} = v_0 2a\theta_s\theta_{az} \\ &= \frac{v_0}{2}\left[8 \ln(2) \frac{\theta_s}{\beta}\right] \frac{\theta_{az}}{\beta} \end{aligned} \tag{4.9}$$

The factor in brackets is Barton's [1, Section 8.2] conical scan error slope, k_s. For a squint angle of 0.5β the error slope is

$$k_s = 8 \ln(2) \times 0.5 = 2.77$$

The other synchronous detector channel after low-pass filtering of $v_{\cos} = v_s \cos(\omega_s t)$ yields the elevation error signal:

$$v_{el} = -v_0 \frac{m}{2} \cos(\alpha)$$

The two error voltage expressions can be rewritten as

$$v_{az} = \frac{v_0}{2} k_s \left(\frac{\theta_{az}}{\beta}\right) \tag{4.10}$$

and

$$v_{el} = \frac{-v_0}{2} k_s \left(\frac{\theta_{el}}{\beta} \right) \tag{4.11}$$

The signs associated with v_{az} and v_{el} are consistent with the way in which angle α is defined in Figure 4.3. The polarity of the servo connection takes these signs into account. Note that v_{az} and v_{el} are proportional to target return strength as embodied in v_0. This target return dependence is removed in a working radar by automatic gain control (Chapter 6) in order to ensure constant tracking servo loop gains, independent of signal strength.

We have just seen that a target produces CONSCAN envelope modulation, the amplitude of which depends on the target's RCS and the angle off boresight. The phase α of the modulation envelope reveals the position of the target on the scan circle (Figure 4.3). The jammer's angle-deception approach is therefore to generate false scan modulation that appears to come from a target at some angle different from the angle of the true target (the vehicle carrying the jammer). The true-target modulation peak occurs at the point on the scan cycle corresponding to the closest approach of the beam to the target ($\phi + \alpha = \pi$ in Figure 4.3). The false target is furthest from the real target, for a given off-boresight distance, if it is dia-

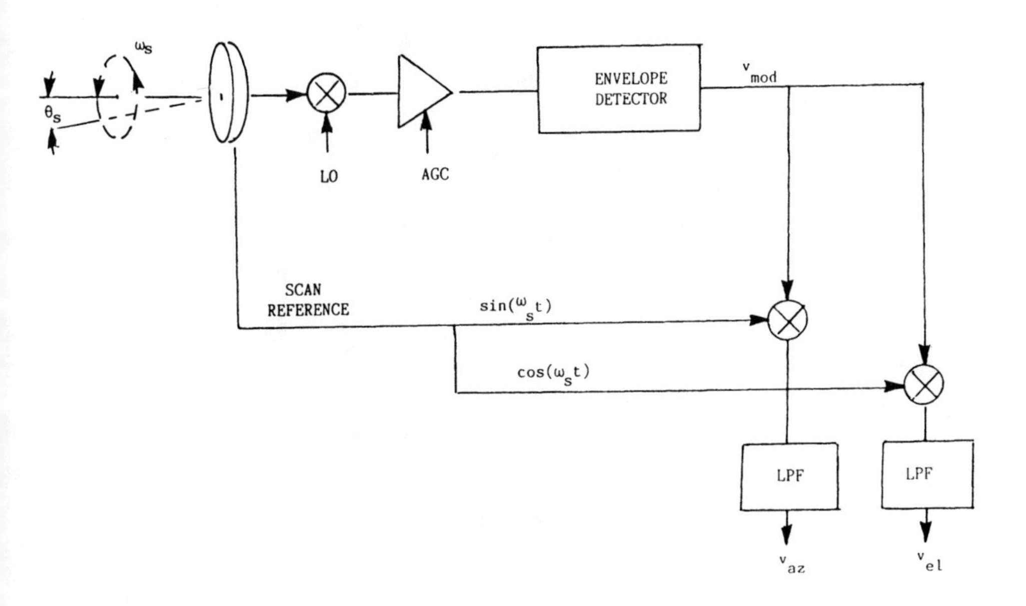

Figure 4.4 CONSCAN error demodulation.

metrically opposite the real target on a line through the boresight axis. To simulate this situation, the false-target modulation peaks must fall in the valleys of the real-target modulation. This form of angle deception is called *inverse-gain modulation,* for it could be produced by making the jammer gain proportional to the reciprocal of the incident illumination level at the jammer. In the presence of this form of angle deception, the automatic tracker shifts its boresight axis to a location at which the sum of the real- and false-target modulation components (that is, the fundamental scan-frequency components) is zero. This null point lies on a line joining the real and false targets, and its location is closer to the stronger of the two.

Many jammers are designed to operate at a fixed (saturation) output level, in which case it is not feasible to impress an arbitrary AM pattern on the jammer output. It is unnecessary to transmit a sinusoidally modulated jammer signal, for the synchronous demodulators (Figure 4.4) retain only the fundamental ω_s component of the modulation. Therefore, inverse-gain angle deception can be implemented by transmitting constant-amplitude jamming bursts timed to coincide with the valleys of the real-target modulation, as indicated in Figure 4.5. The burst timing is synchronized to the CONSCAN cycle by means of an ESM receiver that detects the modulation of the incident radiation from the radar. The fundamental component of the repetitive burst pattern is maximized when the burst duty factor is 50% (square-wave modulation). The duty factor may be lower than 50% to satisfy average-power limitations or to leave part of the jammer power for responding to other threats.

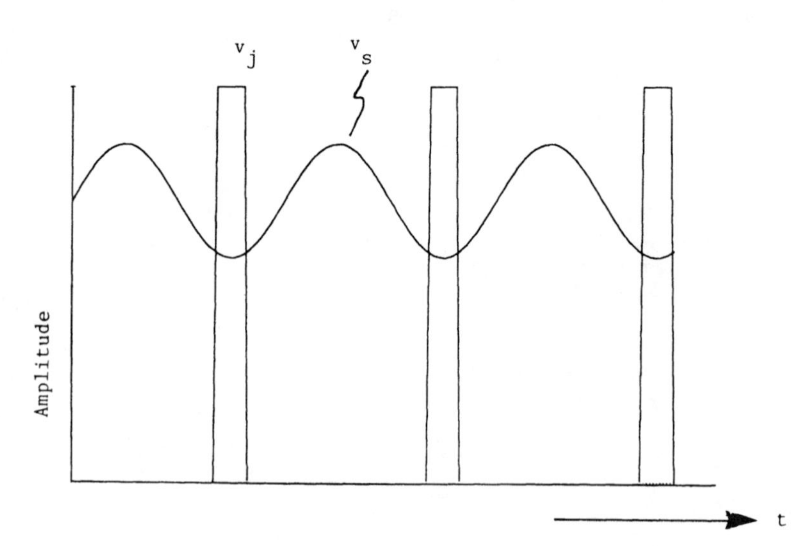

Figure 4.5 Jamming bursts synchronized to CONSCAN cycle.

The effect of inverse-gain angle deception on the CONSCAN tracker cannot be readily analyzed. Although the fundamental scan-frequency component of the repetitive burst pattern can be determined, this pattern is received by the radar only when the radar's boresight axis is on the target (and consequently on the jammer). As soon as a tracking error develops, the received jamming envelope is the transmitted envelope, modulated by the radar's CONSCAN receive pattern. For small errors, the received fundamental component is just the fundamental transmitted component weighted by the dc component of the received pattern. As soon as a large error has developed, the received modulation pattern becomes nonsinusoidal, and fundamental ω_s components arise out of the products of jamming envelope and receiver scan-pattern harmonics, the orders of which differ by 1.

To obtain an approximate measure of inverse-gain jamming effectiveness against CONSCAN, we have developed the following simplified model:

Radar Antenna: Gaussian pattern, single antenna for both transmitting and receiving; β = one-way beamwidth;

Jamming Transmitter: Coherent repeater, jamming signal at radar receiver in phase with target return; burst pattern (constant amplitude), 50% duty factor, synchronized to CONSCAN cycle.

The assumption concerning jammer-target relative phase allows the jamming and signal envelopes to be added algebraically. We computed the fundamental scan-frequency components of the signal envelope, A_{1s}, and of the jamming envelope, A_{1J}, by numerical integration for a range of values of angular tracking error θ_T. The signal component A_{1s} acts to restore the boresight axis to the target (making θ_T zero), and the jamming component tends to drive the tracker off the target. The equilibrium point occurs where $A_{1s} = A_{1J}$.

Figure 4.6 is a plot of A_{1s} and A_{1J} for J/S = 0dB: that is, the received signal and jamming would have had equal amplitudes with the two sources on the peak of the beam (the ratio changes for off-peak locations, because the jamming is weighted by the one-way radar antenna pattern and the target return is weighted by the two-way pattern). The intersection of the two curves is seen to be at θ_T/β = 0.25. This is a stable equilibrium point: $A_{1J} > A_{1s}$ to the left of the intersection, and $A_{1s} > A_{1J}$ to the right of the intersection. Thus, when displaced from this point, the tracker will drive the antenna back toward the equilibrium point.

We determine the effect of J/S on the equilibrium point by shifting the A_{1J} curve vertically relative to the A_{1s} curve. Figure 4.7 is a plot of the normalized tracking error, θ_T/β, as a function of J/S. It is not clear from Figure 4.6 that there is a second intersection at θ_T/β = 2.1, but this represents an unstable equilibrium point. As J/S increases, this second intersection moves toward the first. Finally, when J/S reaches about 15 dB, the stable equilibrium point is lost and a break lock occurs.

The reader must understand that the simplifying assumption concerning jammer-target relative phase results in a pull-off curve that has questionable validity at

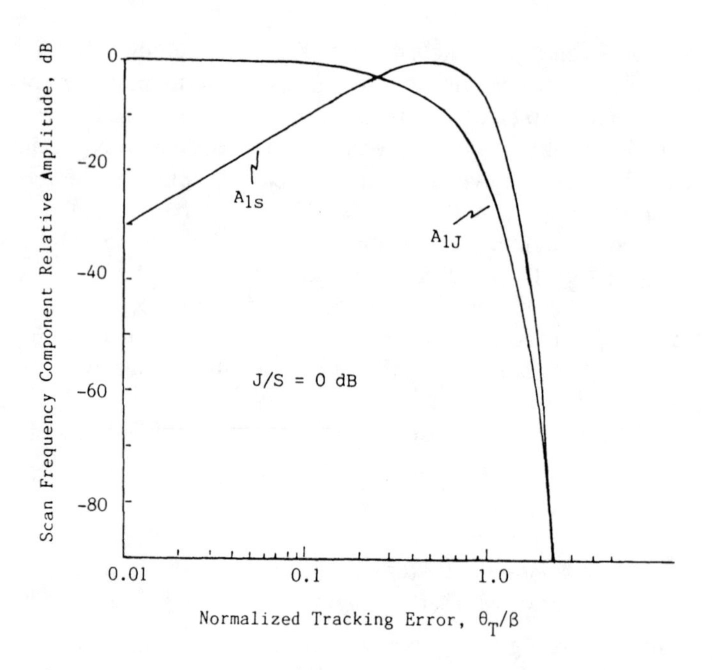

Figure 4.6 Scan-frequency component relative amplitudes for jamming and signal inverse-gain jamming, 50% duty factor.

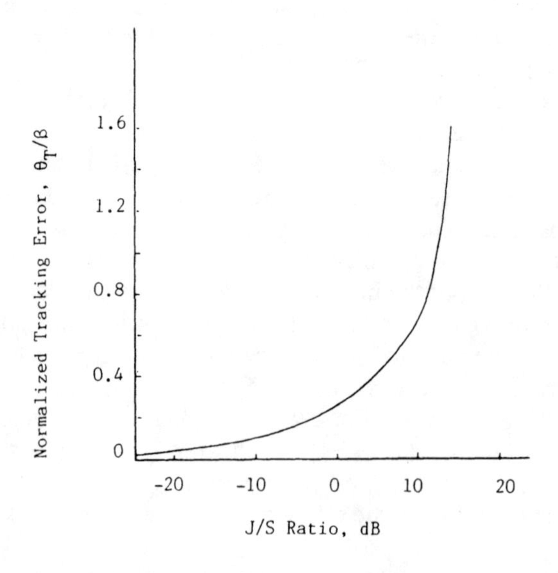

Figure 4.7 Tracking error versus *J/S* Inverse-gain jamming, 50% duty factor.

low *J/S*; however, at high *J/S*, with the jammer dominating the detected envelope, the relative phase is unimportant, so the high *J/S* portion of the pull-off curve should be valid.

A radar-jammer simulation was implemented in accordance with the foregoing description. The tracker was allowed to lock to the target, and jamming was then initiated, causing the tracker to move off the target by an amount depending on *J/S*. Figure 4.8 is a θ_{el} *versus* θ_{az} plot showing the initial lock-on phase in which the tracker was released from the coordinate point (0.01, 0.01) in the absence of jamming (θ_{az}, θ_{el} coordinates are measured in radians).[1] The tracking servo damping was set at $\zeta = 0.5$ (see Appendix A), so the tracker spirals in to a lock on the target at the origin (0.0, 0.0), with about 16% first overshoot on both axes. In Figure 4.9, the tracker was again released at (0.01, 0.01) in the absence of jamming, but, after

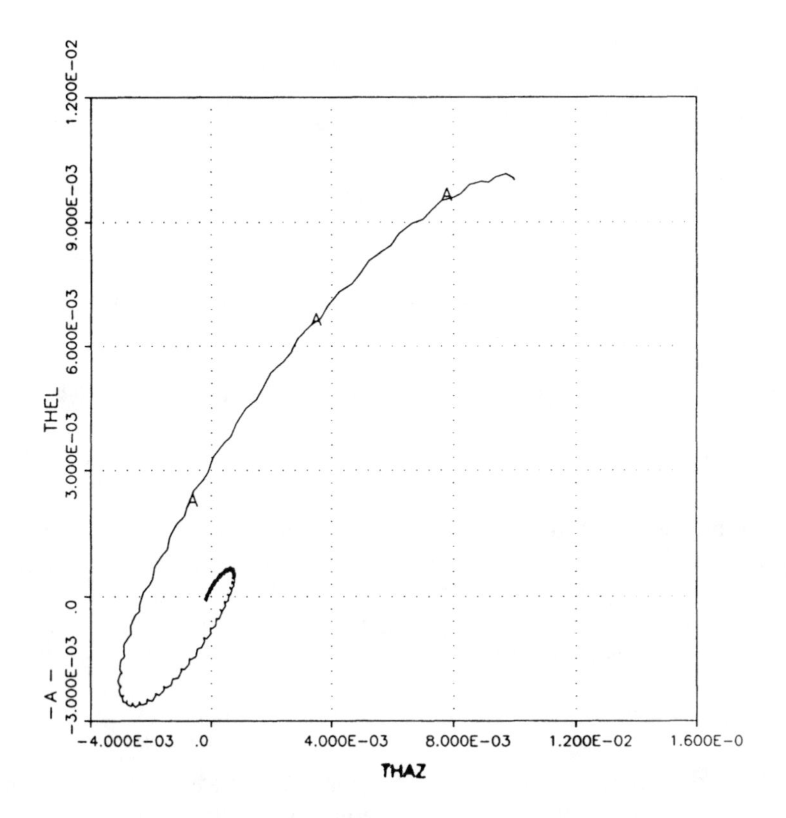

Figure 4.8 Initial lock-on trajectory of CONSCAN tracker (no jamming).

[1]Note the slight residual scan-frequency modulation of θ_{az} and θ_{el} (scan frequency was 30 Hz).

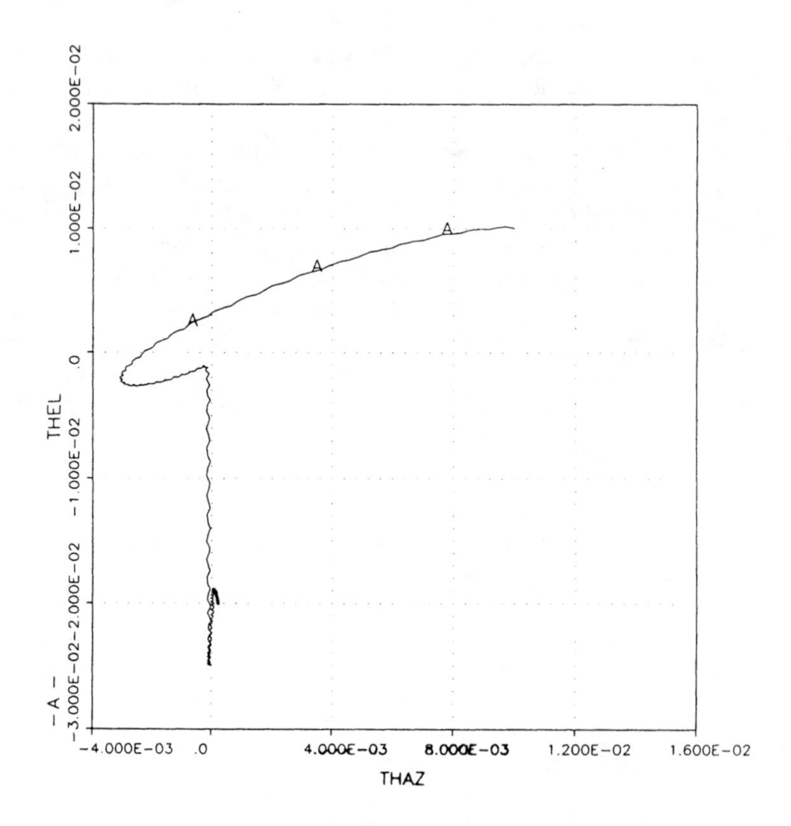

Figure 4.9 Pull-off by inverse-gain jamming as tracker is heading for target ($J/S = -10$ dB), burst duty factor = 50%.

1 s, the jammer was turned on before the tracker had locked on to the target (the coordinates are (0.0, 0.0)). The tracker was then pulled away from the target along the nearly vertical trajectory. J/S was set at -10 dB. The tracker had not completely settled when this plot was terminated (2 s after the jammer was turned on), but it was approaching a value of

$$(\theta_{az}^2 + \theta_{el}^2)^{1/2} \approx 0.02 \text{ rad}$$

This is in good agreement with Figures 4.6 and 4.7. (The direction of the pull-off depends on the position of the tracker boresight axis, relative to the target at the instant the jammer was turned on.) Other simulation runs were conducted at higher J/S values. The pull-off results agreed with Figures 4.6 and 4.7.

4.3 ANGLE DECEPTION AGAINST CONICAL SCAN-ON-RECEIVE-ONLY (COSRO)

In Section 4.2 we assumed that a single antenna was used for both transmission and reception. It was the existence of a scanning transmitting beam that gave the jammer its ability to synchronize its modulation to the scan cycle, producing a false target at a specific angular location relative to the real target. The radar can deny the jammer synchronizing information by illuminating the target with a nonscanning antenna. This radar mode is called *conical scan-on-receive-only* (COSRO). Against COSRO, a jammer is unable to place a false target at a specified location relative to the real target. If the precise value of scan frequency ω_s is known, the jammer can attempt to generate a false target that moves in a circle about the scan axis, thereby causing a time-varying tracking perturbation. The rotation rate of the false target is the difference between ω_s and the jammer modulation frequency ω_j. Ideally, the jammer would set $\omega_j - \omega_s$ to produce frequent perturbations, but the false-target motion must not be too fast for the radar's tracking servos to follow. The offset, $(\omega_j - \omega_s)t$, of the false target takes the place of angle α of the real target in (4.1) and (4.2). Thus, the azimuth coordinate of the false target is

$$\theta_{az} = \theta_j \sin[(\omega_j - \omega_s)t] \tag{4.12}$$

In this equation, an unknown phase has been omitted. The simulated off-boresight angle θ_j corresponds to the θ_T of the real target in (4.7). We noted there that θ_T is proportional to the ratio of the amplitude of the ac (scan-frequency) term to the amplitude of the dc term.[2]

We shall not attempt to relate θ_j to the parameters of the jamming waveform, because the relationship becomes more complex than (4.7) when the target (real or simulated) off-boresight distance becomes an appreciable fraction of a beamwidth. Instead, we simply assume that the jammer signal dominates (high J/S), and that the jammer does generate an unspecified high value of θ_j, perhaps large enough to produce break lock if the false-target motion is not too rapid. The simulated azimuth rate is

$$\dot{\theta}_{az} = \theta_j(\omega_j - \omega_s)\cos[(\omega_j - \omega_s)t] \tag{4.13}$$

Presumably, the jammer programmer would estimate θ_j and set ω_j at a value that keeps the peaks of $\dot{\theta}_{az}$ low enough for the tracker to follow. If $\dot{\theta}_{max}$ is the nominal

[2]The dc term in that equation is essentially independent of θ_T for small values of θ_T.

angular velocity limit of the tracker, then the limit on the false-target scan-frequency offset is

$$\omega_j - \omega_s \leq \frac{\dot{\theta}_{max}}{\theta_j} \tag{4.14}$$

For instance, if $\dot{\theta}_{max}$ is limited to 0.06 rad/s (e.g., the angular rate of a 300-m/s target on a passing course at a range of 5 km) and if $\theta_j = 0.1$ rad, then

$$\omega_j - \omega_s \leq 0.6 \text{ rad/s}$$

for a frequency offset limit of about 0.1 Hz. If the radar's scan frequency is 30 Hz ($\omega_s = 188$ rad/s), this would mean that ω_j must be set to within $\pm 0.3\%$ of ω_s. It is likely that the tolerance on ω_s, as well as the jammer programmer's imperfect knowledge of ω_s, would produce an uncertainty range much greater than $\pm 0.3\%$. Therefore, the only option for the jammer is to modulate its ω_j over the uncertainty range, with the aim of perturbing the tracker whenever ω_j comes close to ω_s. There is an obvious difficulty with this jamming mode, sometimes called a *wobbulation* mode.

We have just given an example in which the jamming is ineffective except when ω_j is very close to ω_s. If the range over which ω_j must be modulated is appreciable, it is clear that the intervals of effectiveness will constitute a small fraction of the total time. Even so, the technique can be useful if it has the potential for creating large perturbations or break locks, provided that the time between these perturbations is not too great. The simplest modulation would be periodic linear sweeps of ω_j across the uncertainty range, $\Delta\omega = \omega_2 - \omega_1$, in a sweep period T_{swp}. The sweep rate is then

$$k = \frac{\Delta\omega}{T_{spw}} = \frac{\omega_2 - \omega_1}{T_{swp}}$$

The warning time—that is, the time between the alert from a radar warning receiver and the actual intercept by a missile—can be quite short, so the jammer should be prepared to induce tracking perturbations spaced by no more than a few seconds in time. With the linear sweep, there is one perturbation per sweep, so T_{swp} is constrained to a few seconds at most. The radar designer can provide an effective countermeasure against this blind-deception technique by providing a sizable scan-frequency adjustment range, thereby forcing a large $\Delta\omega$. The combined low T_{swp} and the high $\Delta\omega$ may result in a sweep rate k so high that the sweep passes through the value ω_s so fast that there is negligible perturbation of the tracker. The false-target scan rate ω_j remains within the bounds set by (4.14) for a time equal to

$$t_p = \frac{\dot{\theta}_{\text{max}}/\theta_j}{k} = \frac{\dot{\theta}_{\text{max}}/\theta_j}{\Delta\omega} T_{\text{swp}}$$

Let us examine the value of the perturbation time, t_p, for the parameters of the following example:

$$\begin{aligned} \dot{\theta}_{\text{max}} &= 0.06 \text{ rad/s;} \\ \theta_j &= 0.1 \text{ rad;} \\ \Delta\omega &= 2\pi \times 10 = 20\pi \text{ rad/s;} \\ T_{\text{swp}} &= 1.0 \text{ s.} \end{aligned}$$

The resulting perturbation time is

$$t_p \approx 10^{-2} \text{ s}$$

If the radar's angle servo bandwidth is B, the duration of the perturbation must be about $1/B$ if it is to be effective. The value of T_p from this example would therefore be effective against a servo with a bandwidth of about 100 Hz. This is perhaps two orders of magnitude greater than might be expected of a CONSCAN tracker. If the servo bandwidth is 1 Hz, the jammer would have to slow its sweep by a factor of 1/100 to produce an effective T_p. This is out of the question, for the perturbations would be so infrequent as to be useless. This is not to say that the jamming would be useless. Even if the deception feature were to fail completely, the jamming could dominate the radar's AGC and, at high J/S, drive the real-target return down to a level too low for accurate tracking.

Given an ESM receiver with a "jog detection" capability, the jammer is able to make an approximate determination of ω_s. The jammer sweeps slowly across the uncertainty band $(\omega_2 - \omega_1)$ looking for a "jog" in the level of incident illumination from the COSRO transmitter. The jog marks the instant at which $\omega_j = \omega_s$. The sweep is stopped at that value of ω_j, and narrow sweeps are initiated about that value. With a much-reduced uncertainty range, T_{swp} can be made short to create frequent tracking perturbations without exceeding an acceptable sweep slope k. This scheme for defeating COSRO will fail against a bistatic COSRO system, such as a semiactive homing missile system using a CONSCAN seeker on the missile and a target illuminator at the launch site.

4.4 ANGLE DECEPTION AGAINST TRACK-WHILE-SCAN (TWS) RADARS

Track-while-scan (TWS) is a combined search and tracking mode that sacrifices the continuous target observation capability of the dedicated tracker in return for the ability to monitor a finite sector of angle space while maintaining tracks on one or more targets moving through that space. The beam is typically a "fan" beam, scanned back and forth in the angle coordinate corresponding to the beam's narrow dimension. Figure 4.10 depicts a vertical fan beam of width β_{az} scanning an azimuth sector of width $\Delta\theta_{az}$. This radar provides range and azimuth (R, α) coordinate information on targets detected anywhere within the vertical fan beam.

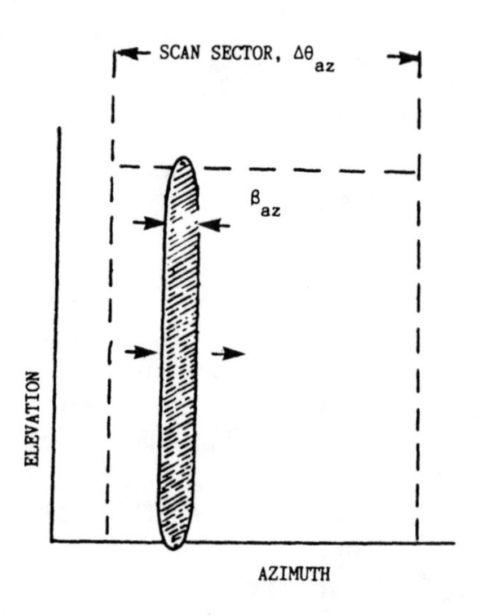

Figure 4.10 TWS scan in azimuth.

Figure 4.11 shows a range-azimuth (B-scope) display with three target detections. If all three are to be tracked, a range tracker must be assigned to each, and an angle tracker must be assigned to each range tracker. The range tracker may employ any of the techniques discussed in Chapter 3, but it differs from those in that target returns are available on an intermittent basis (i.e., only during the beam's transit across the target). Between target updates, the tracker must coast. Hence, TWS radars associated with weapons systems generally scan at high rates (many scans per second). Once a target is acquired, the range gate is enabled only during the time the target is within the angle gate, and the angle tracker receives returns only from its associated range gate. As the beam scans across the target, a burst of return pulses is received with an amplitude envelope corresponding to the beam pattern. The target return burst, after processing by a boxcar circuit, is a stair-step replica of the beam pattern of the sort depicted in Figure 4.12.

The angle tracker is typically a split-gate type, identical in concept to the split-gate range tracker but with range delay time replaced by azimuth scan time. The members of the split-gate pair for an azimuth tracker are called the right and left

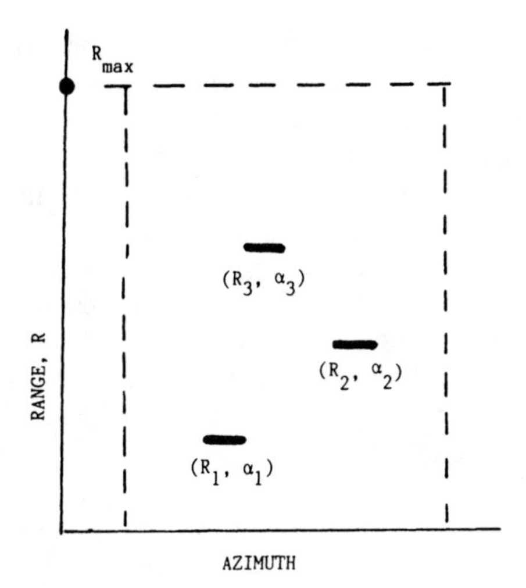

Figure 4.11 Azimuth and range display.

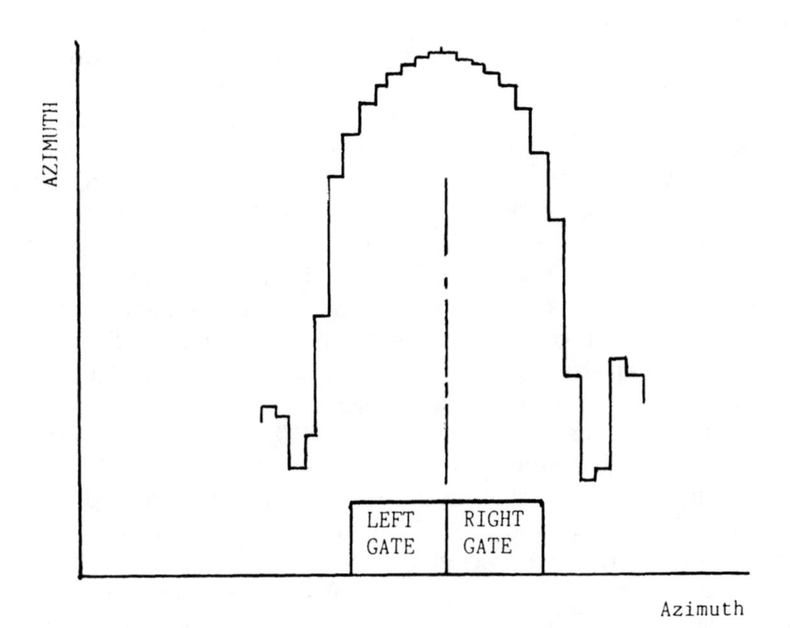

Figure 4.12 TWS angle gate centered on target.

gates (Figure 4.12). Each gate integrates its share of the target return envelope to obtain a voltage-time area value. When the gate is centered on the target, the areas are equal and the error signal (right minus left) is zero. The azimuth servo corrects the timing of the split gate within the scan cycle as necessary to zero out the error. There may be, in addition, a pedestal servo that mechanically steers the antenna to keep the radar's field of view centered on the target.

Generally the TWS radar acquires targets that are handed off to it by an associated surveillance radar, but some TWS radars are capable of performing their own surveillance. One way in which they do this is to stop the rapid TWS scan and switch to a slower mechanical scan of the pedestal over a larger surveillance sector (or even a 360° scan). This is likely to be only a backup surveillance mode, for the range of the TWS fire control radar is generally shorter than optimum for surveillance.

In some applications, only two-dimensional (range and azimuth) information is required, but in fire control systems involving command guidance or for aiming antiaircraft artillery, three-dimensional (range, azimuth, and elevation) target coordinates are required. The elevation information is obtained from a second TWS system by using a horizontal fan beam of width β_{el} scanning elevation sector $\Delta\theta_{el}$. The two TWS systems (their coverage areas coincide) are, in effect, independent radars. Target range, which is common to both systems, is used for associating azimuth and elevation coordinate information with the proper target.

When the scanning fan beam is used by the radar for both transmission and reception, a target aircraft equipped with an ESM receiver is able to detect the passage of the beam across the target and to coordinate its jamming modulation with the scan cycle. *Angle-gate pull-off* (AGPO) tactics against the TWS radar are analogous to range-gate pull-off except that time within the scan cycle replaces range delay time. The AGPO deception waveform is typically a burst of transponder or repeater pulses of duration equal to the beam dwell time and, in that respect, resembles target return pulses, although they are likely to be transmitted at constant amplitude. The burst is initially timed to coincide with the passage of the transmitted beam across the target. This timing ensures that the burst will enter the TWS angle gate, and the fact that the pulses are generated by a transponder or a repeater ensures their passage through the radar's range gate.

The AGPO process may be preceded by RGPO, thereby luring the range gate to the false-target burst, separated in range from the real target. If the RGPO succeeds, J/S in the radar receiver is infinite, ensuring that the AGPO deception will succeed. If RGPO fails, subsequent AGPO is certain to fail. The AGPO procedure consists of slowly shifting the deception burst timing within the TWS scan cycle away from the position of the target return. If J/S is sufficiently high and if the deceptive angle motion of the burst is not too fast, the TWS angle gate will be lured from the target. Once the false-target position has been shifted well off the real target, the transponder or repeater transmissions are triggered from sidelobe emissions from the TWS radar. The alternative to the AGPO burst of pulses is a continuous burst (for instance a noise burst) with the appropriate envelope. The average power of the continuous burst is much higher, and it provides no opportunity for RGPO.

The J/S requirements for AGPO are essentially the same as for RGPO because the tracking circuits are of the same form. In AGPO there is no countermeasure corresponding to the leading-edge tracking tactic used against RGPO. The TWS scan cycle is repetitive, so there is nothing to hinder the placement of the false-target burst of AGPO precisely on top of the target return envelope or on either side of the target envelope.

A simple AGPO simulation was run in which a repeater jammer emitted bursts of duration equal to half the main-lobe width, β, of the TWS beam. With jamming arriving in phase with the target return, a tracking perturbation of magnitude $\beta/2$ was achieved with $J/S \approx 3.5$ dB. With the jamming arriving with 180° phase shift relative to the target return, a J/S of approximately 9 dB was required to produce the same tracking error.

As we shall see in the next section, a very effective countermeasure against TWS angle deception is to deny the jammer the ability to synchronize with the radar's scan cycle. This is accomplished by going to a SORO mode in which a non-scanning transmitting beam illuminates the entire field of view. The cost, in addition to the cost of a separate transmitting antenna, is a transmitting gain sacrifice equal to the ratio of the solid angle of the fan beam to the solid angle of the non-scanning transmitted beam:

$$\text{SORO gain sacrifice} = \frac{\beta}{\Delta\theta}$$

For instance, if $\beta = 2°$ and $\Delta\theta = 20°$, the sacrifice amounts to 10 dB.

4.5 ANGLE DECEPTION AGAINST SORO-TWS[3]

SORO operation leaves the jammer aboard a target aircraft entirely ignorant of the timing of the radar's TWS scan cycle. AGPO attempts must be made blindly, using a jammer modulation pattern designed to move the false target (the jammer burst) across the entire TWS scan sector. The objective is to cause the false target to traverse the TWS angle gate, wherever it may lie, and lure the gate off the target. Even if the AGPO maneuver fails to dislodge the gate, it is likely to cause some angle

[3]This section also applies to AGPO against a TWS radar with transmitting and receiving beams scanning together, if the jammer lacks the capability to make use of scan pattern information. In fact, a SORO mode is unlikely when a large sector is covered by the TWS scan.

perturbation of the tracker. Because there is no easy way[4] to determine the degree of success, the jammer modulation pattern must be repeated often, thereby generating repeated perturbations or repeated break locks. If the perturbations are large enough and frequent enough, the fire control system may be deterred from launching a missile or command guidance after launch may be disrupted.

This form of jamming can be effected with one jamming burst per TWS scan cycle. If the TWS scan frequency is known, the burst rate can be set slightly higher or slightly lower than the scan rate, causing the false target to move forward or backward across the scan sector by small steps, $\delta\theta$, per scan. The burst duration would be, typically, about the same as the angle-gate duration. As the burst enters the gate from the left, the signal centroid is shifted to the left, causing a tracking perturbation in that direction. As the burst moves on through the gate, it may carry the gate with it, but failing that it creates a tracking perturbation to the right as it leaves.

If the angle-gate width is equal to the beamwidth β, the gate occupies a fraction $\beta/\Delta\theta$ of the angle space, where $\Delta\theta$ is the width of the scanned sector. With a single burst per scan cycle, the slowly moving burst is perturbing the gate for approximately[5] that same fraction of the time. The passage of false targets through the angle gate can be made n_B times as frequent by generating n_B bursts per scan. The burst spacing must be great enough to permit one burst to leave the trailing edge of the gate before the next burst crosses the leading edge. Figure 4.13 depicts a jamming pattern (left side of the plot) in which there are four bursts per scan. The scan is represented by the sawtooth pattern (linear scan with zero retrace time), and the burst locations on the scans are indicated by dots. The angle-gate boundaries are indicated by the dashed lines. The burst pattern is seen progressing through the gate in about eight scan cycles. The next burst set is ready to enter the gate at about the time the first set leaves.[6]

For this form of jamming to be effective, the false target must move through the gate at a rate slow enough for the tracking servo to respond, either breaking lock or generating tracking perturbations of significant magnitude. If the transit is too fast, the tracking perturbation is negligible. If the TWS scan frequency is accurately known, there is no difficulty in setting the burst rate to create the desired false-target transit speed. It is advantageous to the radar that it not have a single

[4]Jog detection may reveal a track perturbation only if the perturbation is a significant fraction of the transmitted beamwidth $\Delta\theta$. (See Section 4.3, where job detection is discussed in connection with ECM against COSRO.)

[5]Actually, if the burst duration is equal to the gate duration, some portion of the burst is in the gate for a fraction, $2\beta/\Delta\theta$, of the time.

[6]This burst spacing is somewhat too close if the burst width is equal to the gate width. However, the gate is likely to occupy a considerably smaller fraction of the scan than is depicted here.

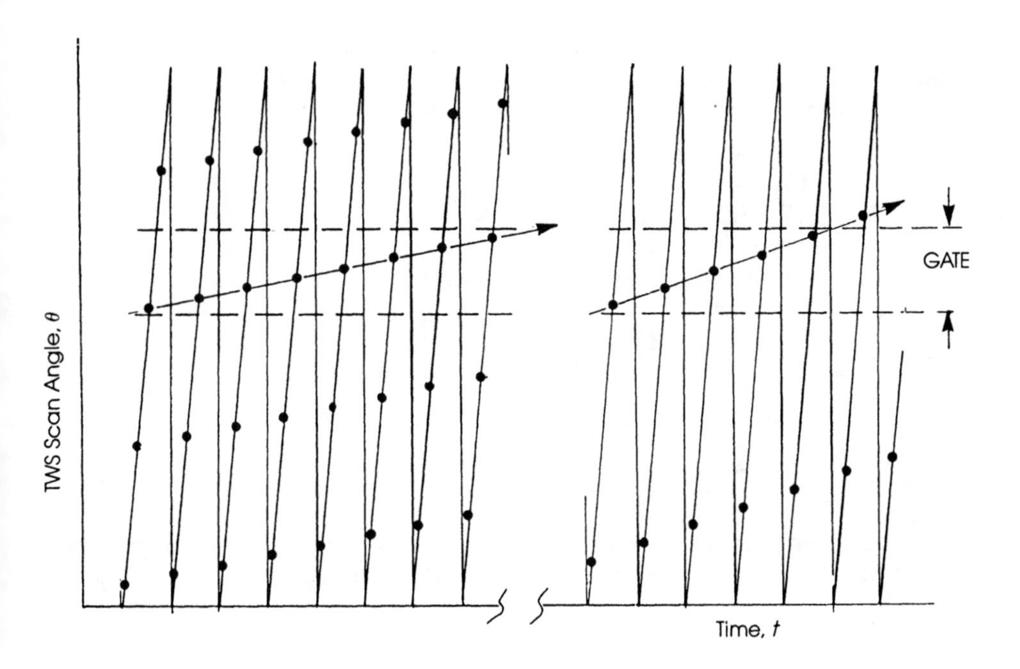

Figure 4.13 Passage of jamming bursts through TWS angle gate.

scan frequency (its value may be learned by the enemy) but rather an ability to set the scan frequency anywhere within a finite adjustment range. Clearly, if the jammer burst rate is based on an erroneous estimate of TWS scan frequency, the transit rate of false targets through the angle gate can be very different from the intended rate.

The only strategy for the jammer operating against a TWS radar with an unknown scan frequency is to vary the burst rate over a range of values calculated to be effective at least part of the time. The range of burst rates would be selected to match the known (or estimated) radar scan-rate adjustment range. This mode of jamming is sometimes called *wobbulation.* The simplest jammer program for attacking the SORO-TWS radar would be a linear sweep of the burst rate across the range of values computed from the estimated scan-rate adjustment range. Figure 4.13 is intended to imply such a linear sweep of burst rate, starting with close spacing (four bursts per scan) at the left of the plot and increasing (to only two bursts per scan) toward the right side of the plot. One might object to a repetitive sweep because a very intelligent tracker might identify the periodicity of the sweep and coast through the periods of tracking perturbation.

The following discussion indicates some of the compromises that must be made in the design of a jamming program for use against a SORO-TWS radar with adjustable scan frequency. The notation is as follows:

β = TWS antenna beamwidth;
$\Delta\theta$ = TWS scan sector;
T_s = TWS scan period (assume saw-toothed scan);
B = angle servo bandwidth;
n_B = number of jammer bursts per scan;
T_B = burst spacing;
ω_j = angular rate of burst pattern across angle sector;
$\omega_{T\max}$ = maximum expected angular rate of real targets.

The bursts progress across the scan period in steps of δt per scan:

$$\delta t = n_B T_B - T_s$$

This progression is in the positive (increasing θ) direction or the reverse direction, depending on the sign of δt. A time step of size δt translates to an angle step $\delta\theta$ given by

$$\delta\theta = \Omega\delta t$$

where Ω is the actual rate at which the TWS beam scans:

$$\Omega = \frac{\Delta\theta}{T_s}$$

Thus, the angular rate of transit of false targets across the scan sector is

$$\omega_j = \frac{\delta\theta}{T_s}$$

The portion of the jamming program cycle (e.g., linear sweep) during which large tracking perturbations and potential break locks can be expected is that portion for which:

$$|\omega_j| \leq \omega_{T\max}$$

This expectation is true not only because the tracking servo-response time (and servo bandwidth) can be expected to accommodate real targets with some maximum angular rate, $\omega_{T\max}$, but also because an intelligent tracker may deliberately reject targets with unrealistic rates.

The jammer programmer must be concerned with the following questions:

- For what fraction of the sweep is $\omega_j \leq \omega_{T\max}$?
- How large is the potential angle perturbation?
- How often do perturbations occur?

Obviously, speeding up the sweep rate will increase the number of false-target transits per second, but the transits may become too fast to be effective.

Figure 4.14 depicts one cycle of a jammer program in which the burst spacing is swept linearly over the range T_{B1} to T_{B2} in a period T_{swp}. The sweep slope is

$$k = \frac{T_{B2} - T_{B1}}{T_{\text{swp}}}$$

At the point labeled $t = 0$, the burst rate reaches the value

$$T_B = T_{B0} = \frac{T_S}{n_B}$$

at which point the burst progression across the TWS scan sector momentarily comes to a stop and reverses direction. It is in the vicinity of this point that the false-target's angular rate corresponds to the angular rates of real targets. The region between $-t_1$ and $+t_1$ is the region in which $|\omega_j| \leq \omega_{T\max}$. There, T_B is described by

$$T_B = T_{B0} + kt$$

At the edge of that region ($t = t_1$):

$$\delta t = n_B T_B - T_s = n_B(T_{B0} + kt_1) - T_s = n_B k t_1$$

and the corresponding false-target transit rate is

$$|\omega_j| = \omega_{T\max} = \frac{\Omega \delta t}{T_s} = \frac{\Omega n_B k t_1}{T_s}$$

Therefore,

$$t_1 = \frac{\omega_{T\max} T_s}{\Omega n_B k}$$

The angular perturbation of the false target during the time interval $0 < t < t_1$ is

$$\theta_P = \int_0^{t_1} \omega_j \, dt = \frac{\Omega n_B k}{T_s} \int_0^{t_1} t \, dt = \frac{\Omega n_B k t_1^2}{2T_s}$$

$$= \frac{\omega_{T\max}^2 T_s}{2\Omega n_B k} = \frac{\omega_{T\max} t_1}{2}$$

There is a mirror image movement on the other side of $t = 0$.

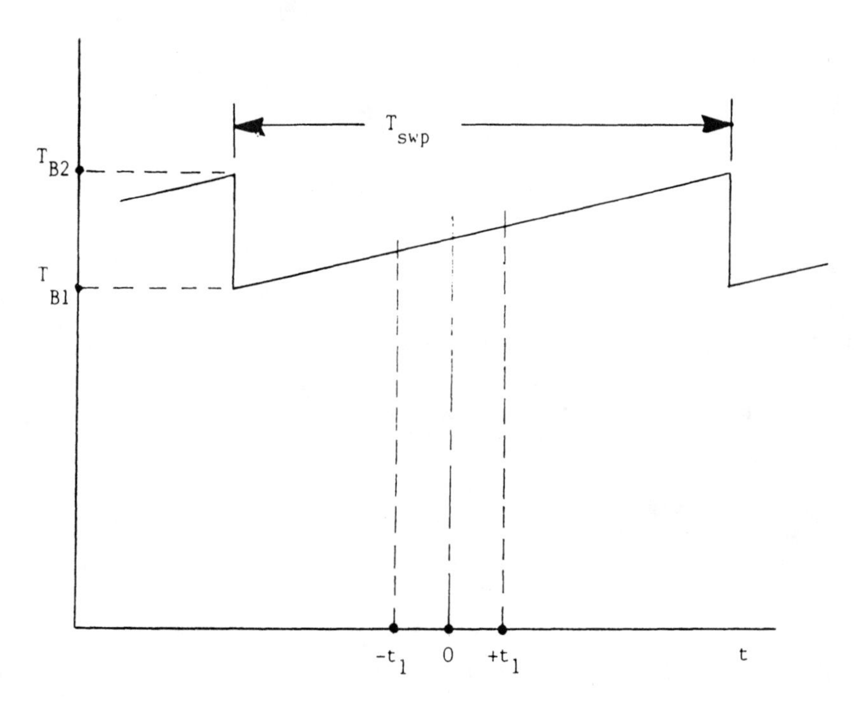

Figure 4.14 Linearly swept burst rate.

The following conditions are necessary for the sweep to be effective against the TWS radar:

1. The perturbation duration t_1 must be long enough for the angle servo to respond. If B is the servo bandwidth, then the requirement is

$$t_1 \geq \frac{1}{B}$$

2. The magnitude θ_p of the perturbation must be a significant fraction of a beamwidth.

3. Perturbations must be frequent. There is one perturbation per sweep. A reasonable requirement might be

$$T_{\text{swp}} \leq 3 \text{ s}$$

4. The jammer burst sequence must be within the TWS angle gate during the interval, $2t_1$, for which $|\omega_j| \leq \omega_{T\max}$.

This final condition is ensured if, as in the left side of Figure 4.13, a new burst sequence is ready to enter the gate as soon as its predecessor leaves. With wide burst spacing (small n_B) there may be no bursts within the gate during this crucial portion of the sweep.

Example:

$$\begin{aligned} \Delta\theta &= 20° \text{ scan sector;} \\ \beta &= 2° \text{ beamwidth;} \\ B &= 2.0\text{-Hz tracker servo bandwidth;} \\ T_s &= 0.05\text{-s nominal scan period;} \\ n_B &= 4 = \text{nominal number of bursts/scan;} \\ \omega_{T\max} &= 0.06 \text{ rad/s.} \end{aligned}$$

The value of $\omega_{T\max}$ corresponds, for instance, to a 300-m/s target on a passing course at a minimum passing range of 5 km. The value of Ω is

$$\Omega = \frac{\Delta\theta}{T_s} = 400°/\text{s} = 6.98 \text{ rad/s}$$

Let us suppose that the uncertainty as to the precise scan period is $\pm 5\%$. This means that $n_B T_B$ must be swept over a 10% range centered on the nominal value of T_s:

$$4(T_{B2} - T_{B1}) = \frac{T_s}{10} = 5 \times 10^{-3} \text{ s}$$

Let us set the sweep period at

$$T_{\text{swp}} = 3.0 \text{ s}$$

The resultant sweep rate is

$$k = \frac{T_{B2} - T_{B1}}{T_{\text{swp}}} = \frac{5 \times 10^{-3}}{12} = 4.17 \times 10^{-4}$$

Then,

$$t_1 = \frac{\omega_{T\max} T_s}{\Omega n_B k} = 0.26 \text{ s}$$

$$\theta_P = \frac{\omega_{T\max} t_1}{2} = 7.73 \times 10^{-3} \text{ rad} = 0.44°$$

and

$$\frac{\theta_P}{\beta} = 0.22$$

Apparently, the sweep will, at best, be marginally effective. The potential perturbation is less than a quarter of a beamwidth. We can achieve this potential only if $t_1 \approx 1/B$, but we find that

$$t_1 B = 0.26 \times 2 = 0.52$$

so the actual perturbation will be less than θ_P, perhaps no more than a tenth of a beamwidth. Note that, in this example, there is no way to make the jamming more effective. Both t_1 and θ_P are smaller than we like by a factor of at least ½. We can double those values by doubling T_{swp}, but this would double the time between perturbations, which is already longer than we want.

If a jog detector (see Section 4.3) could be used to detect the instant of passage through the TWS scan rate, the sweep could be reduced in width and T_{swp} could be reduced. It is unlikely that the perturbation of the broad transmitting beam could be detected, for even if the full $\theta_P = 0.22°$ were achieved this would only be about 1/100 of the $\Delta\theta = 20°$ transmitting beamwidth.

4.6 ANGLE DECEPTION AGAINST MONOPULSE

Sequential lobing in a single plane uses a single antenna lobe the position of which alternates between the squint angles, $\pm\theta_s$, on either side of the boresight. In monopulse tracking the two lobes are present simultaneously. If $f_1(\theta)$ is the voltage pattern of the left lobe and $f_2(\theta)$ is the pattern of the right lobe (see Appendix B), the difference pattern is

$$\Delta(\theta) = f_1(\theta) - f_2(\theta)$$

and the sum pattern is

$$\Sigma(\theta) = f_1(\theta) + f_2(\theta)$$

A target of amplitude v_T at angle θ_T produces difference and sum channel signals $v_T\Delta(\theta_T)$ and $v_T\Sigma(\theta_T)$. The monopulse error signal V_ϵ is the ratio of the difference signal to the sum signal:

$$V_\epsilon(\theta_T) = \frac{v_T\Delta(\theta_T)}{v_T\Sigma\,(\theta_T)}\cos(\phi) = \frac{\Delta(\theta_T)}{\Sigma\,(\theta_T)}\cos(\phi)$$

where ϕ is the phase angle between the Δ and Σ channel voltages. The channel phases are adjusted to make $\phi = 0°$ or $180°$, depending on the sense of the error, so the sign of $\cos(\phi)$ corresponds to the polarity or sense of error θ_T. Because the Δ and Σ channel signals are formed essentially instantaneously, the error signal is independent of target amplitude. Neither target amplitude scintillation nor amplitude-modulated jamming from a jammer aboard the target vehicle can induce tracking errors. In fact, such a jammer serves as a beacon, providing the monopulse system with a strong signal for angle tracking.

Various jamming schemes have been devised for exploiting imperfections in monopulse signal processing, such as phase and amplitude unbalances in the receiver signal paths. The effectiveness of such schemes depends on the level of imperfection present. Generally, against a well-designed, properly maintained monopulse radar, the tracking errors induced by such jamming are small. Two jamming schemes, CROSSEYE and CROSS-POL, attack the monopulse tracker by presenting the monopulse antenna with a wavefront different from the one for which it is designed.[7]

4.6.1 CROSSEYE

A direction finder or an angle tracker bases its measurement on the orientation of the incident plane wavefront. The source of the wave is presumed to lie on the normal to the wavefront. It is well known that a composite target made up of several scatters exhibits glint effects, causing the apparent line-of-sight direction to the target to fluctuate. CROSSEYE is a scheme that carries glint to the extreme by creating two apparent strong scatterers aboard the target vehicle. In Figure 4.15 the aircraft carries a pair of jammers, J_1 and J_2, mounted on the wingtips. The aircraft is approaching a radar at range R. The jammers subtend angles δ from the line of sight, given by

$$\delta = \frac{d}{2R}$$

[7] Although CROSSEYE and CROSS-POL jamming will induce angle errors in nonmonopulse trackers, the effects cannot be assumed to be identical to the effects on monopulse.

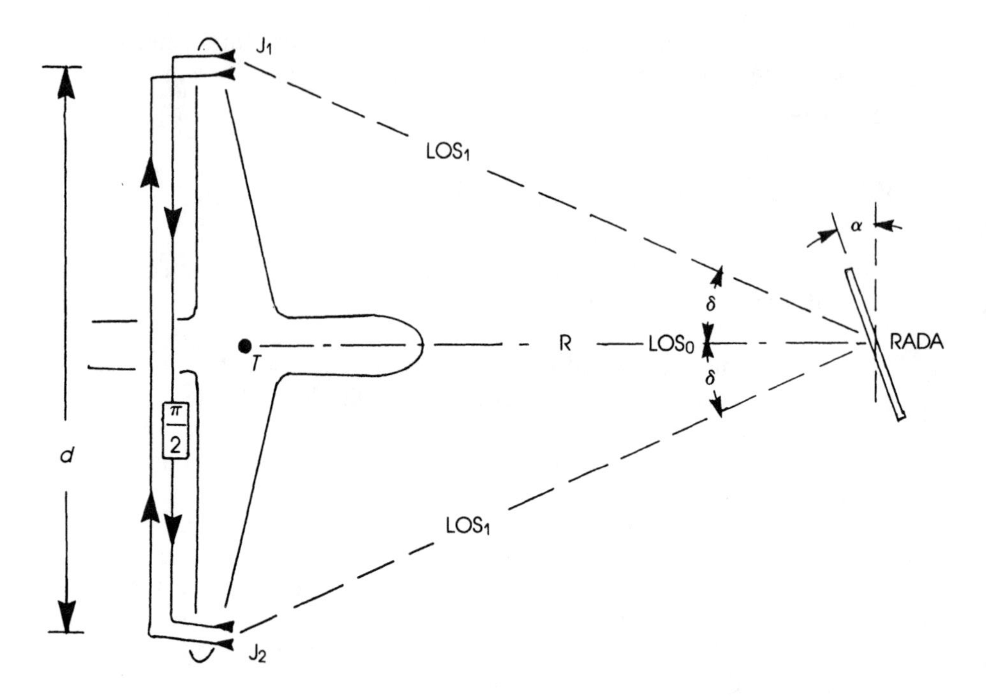

Figure 4.15 CROSSEYE geometry.

where d is the jammer separation. The two jammers are repeaters, but J_1 repeats the radar pulse received by J_2 and J_2 repeats the pulse received by J_1. The lengths of the two cross-coupling cables between J_1 and J_2 are carefully matched, but one path has an extra phase lag of 180°. We shall see later the reason for the cross-linking.

For the geometry of Figure 4.15, the line-of-sight paths from the radar to the two jammers, LOS_1 and LOS_2, are of equal length, so the phases of the radar pulse received at the two jammers are identical. With equal gains, the repeaters emit identical replicas of the radar pulse except for the 180° phase difference. The resultant radiation pattern of the two-element array is a multilobed interferometer pattern with a central null precisely along LOS_0, the line of sight to the center of the radar antenna aperture.

The far-field pattern is of the form:

$$E(\phi) = E_0 \sin\left(\frac{\pi d}{\lambda}\sin(\phi)\right)$$

where ϕ is the angle of the far-field measurement point relative to the line LOS_0, λ is the wavelength, and E_0 is the field strength at the peaks of the lobes. We are interested only in the field across the radar antenna aperture. Because the aperture subtends a very small angle, we can replace $\sin(\phi)$ by x/R, where x is the distance along the radar antenna aperture, measured from its center, point 0. The field across the aperture is, therefore,

$$E(x) = E_0 \sin\left(\frac{\pi d x}{\lambda R}\right) \approx E_0 \frac{\pi d x}{\lambda R}$$

Thus, the CROSSEYE field differs drastically from the field that would be received from a point scatterer or an emitter located at point T, the center of the target vehicle. The CROSSEYE field has odd symmetry relative to the aperture center, whereas the plane wave from the single point source has even symmetry. The odd symmetry occurs because the radar antenna is centered on the central null of the CROSSEYE interferometer pattern the lobes of which alternate in polarity.[8] Consequently the CROSSEYE emission produces a Σ channel null and a Δ channel maximum when the radar antenna is normal to LOS_0. This is just the opposite of the response to a point target on boresight. In the absence of contamination, such as target backscatter or clutter returns, the CROSSEYE voltages developed in the monopulse Δ and Σ channels result in an error voltage that drives the monopulse antenna's boresight axis off the target vehicle. The tracker comes to rest at one of the stable CROSSEYE nulls, the locations of which we shall investigate.

Before doing so, we will explain the effect of cross-linking the repeaters. If the cross-links were absent and each jammer simply repeated its own received radar pulse but with 180° greater phase lag in one repeater path, a yaw motion of the aircraft would cause a yaw rotation of the multilobed interferometer pattern (the pattern would rotate twice as far as the aircraft). There would be no way to maintain a null at the radar antenna. We can easily show that cross-linking the jammers cancels the yaw-induced phase lags, thereby maintaining the null on the radar antenna.

The response of monopulse to CROSSEYE can be analyzed from either of two viewpoints: (1) the integration of the CROSSEYE illumination across the radar receive aperture; or (2) the summation of the two far-field CROSSEYE emitter contributions, weighted by the patterns of the pair of squinted beams from which the monopulse Δ and Σ patterns are formed. We shall take the latter approach. For simplicity we use an elemental $(\sin x)/x$ beam:

$$f(\theta) = \frac{\sin(k\theta)}{k\theta}$$

so the one-way voltage patterns of the two squinted beams are

[8] The relative phase received at a given range in adjacent lobes is 180°.

$$f_1(\theta) = \frac{\sin[k(\theta - \theta_s)]}{k(\theta - \theta_s)}$$

and

$$f_2(\theta) = \frac{\sin[k(\theta + \theta_s)]}{k(\theta + \theta_s)}$$

where θ is the off-boresight angle of the far-field measurement point, and θ_s is the beam squint angle. The constant k is

$$k = \frac{2.78}{\beta_1}$$

where β_1 is the 3-dB width of the elemental beam. We shall be summing up the contributions of jammers J_1 and J_2 and target T to the voltage received by each beam. The boresight axis of the radar antenna is steered to the angle α relative to LOS_0 (Figure 4.15). The angles of arrival of the wavefronts from each of the sources are then:

for J_1, $\theta = \theta_1 = \alpha - \delta$;

for J_2, $\theta = \theta_2 = \alpha + \delta$;

for T, $\theta = \theta_0 = \alpha$.

We can determine the potential of pure CROSSEYE jamming for perturbing a monopulse tracker by locating the stable CROSSEYE tracking nulls. These are the values of antenna pointing angle, α, for which Δ goes to zero in the presence of CROSSEYE illumination, and Σ is of the correct polarity to produce a restoring torque when the antenna is perturbed off the Δ null.

We can find the CROSSEYE tracking nulls by computation or by simulation. Figure 4.16 gives a simulation diagram for determining the monopulse tracker response to CROSSEYE experimentally. The inputs E_1 and E_2 represent the field contributions of jammers J_1 and J_2 at the radar antenna. They arrive from angles θ_1 and θ_2, respectively, and are in phase opposition (as are J_1 and J_2). Target return E_s was arbitrarily set in phase with E_1. We defined

$$J/S = \left(\frac{E_1}{E_s}\right)^2$$

We made the amplitude E_s adjustable to permit simulation with any desired J/S. A simple antenna servo was modeled with a damping ratio of 50%. The simulation results are plotted in Figure 4.17. The jammers were initially turned off,

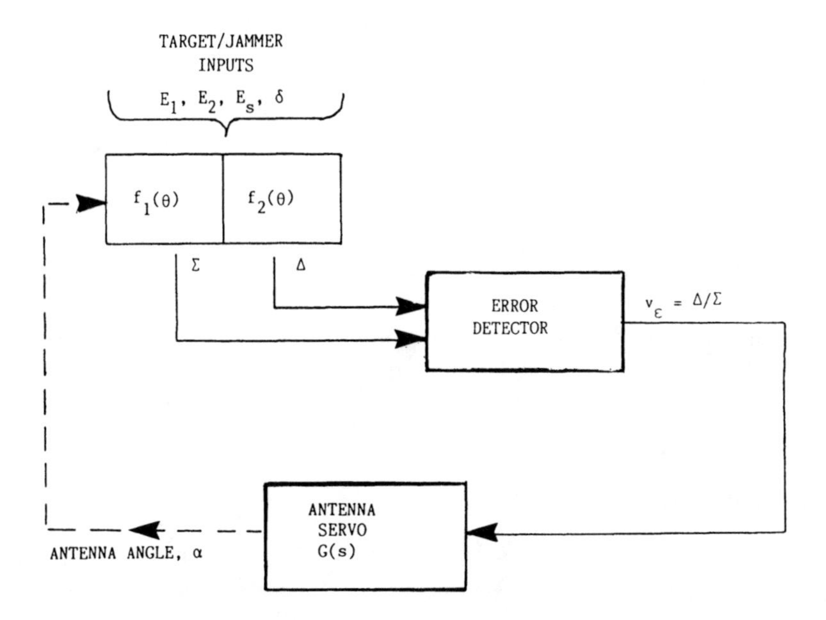

Figure 4.16 Simulation of monopulse response to CROSSEYE.

and the tracker was allowed to lock onto the target return ($\alpha = 0$). As J/S increases, the tracker is pulled off the target toward the first CROSSEYE tracking null at $\alpha_1 = 0.62\beta$, where $\beta = 1.22\beta_1$ is the 3-dB width of the Σ beam. This value of β corresponds to $\theta_s = 0.5\beta_1$ used in the simulation. To locate the next CROSSEYE tracking null, we again turned off the jammers and allowed the tracker to lock onto the target on the first Δ sidelobe null at $\alpha_{SL} = 1.39\beta$, after which J/S was gradually increased, pulling the tracker to the second CROSSEYE tracking null at $\alpha_2 = 1.79\beta$. Note that it takes $J/S = 45$ dB to pull halfway to the CROSSEYE tracking null and $J/S = 65$ dB to get 90% of the way. These high required ratios are just one of the difficulties of implementing CROSSEYE.

We should not be surprised to learn that high J/S is required. With the CROSSEYE illumination null centered on the radar antenna aperture, even the edges of the aperture are far down in the null of the lobe pattern. The average field strength across the aperture is, approximately,

$$\overline{E} = \frac{\pi d W}{4 R \lambda} E_0$$

where E_0 is the field strength at the peaks of the lobe pattern, d is the CROSSEYE emitter spacing, W is the radar antenna aperture width, and R is the range. For d

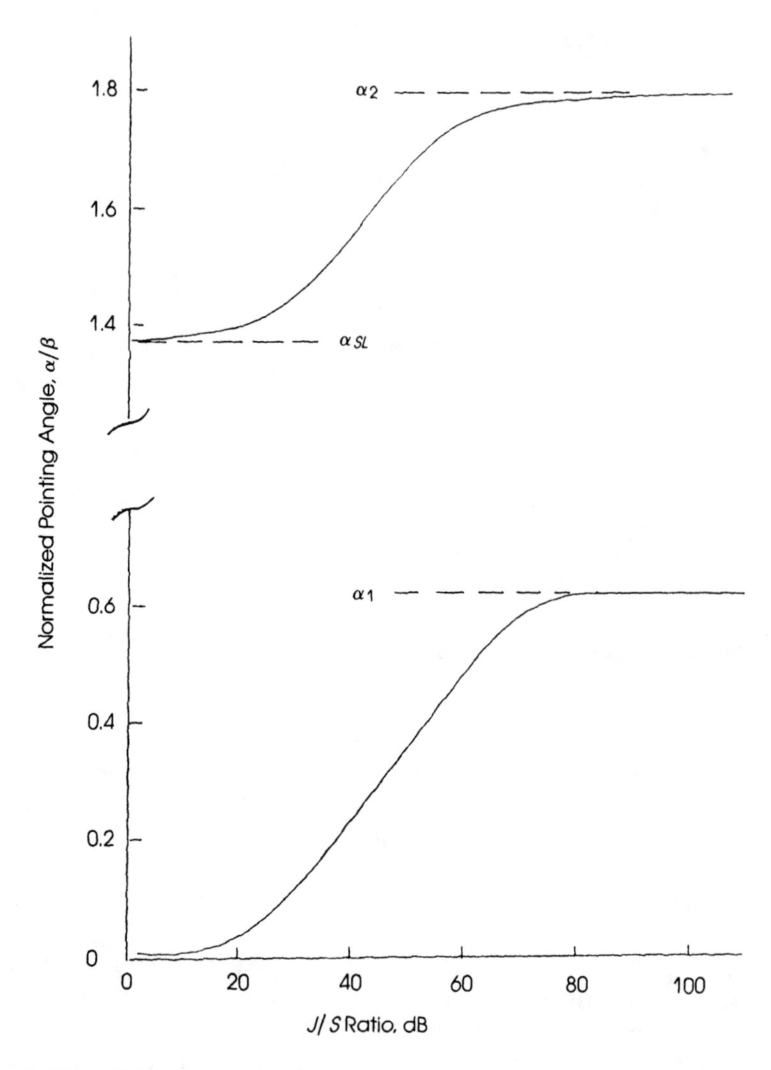

Figure 4.17 CROSSEYE pointing error versus J/S.

= 10 m, W = 1.0 m, λ = 0.03 m, and R = 5 km, this average field strength is about 26 dB below the strength on the lobe peaks. The jammer must not only overcome this null attenuation but it must also overcome the "pull" of the target. We can visualize this latter factor through the simple model described by Figure 4.18. The CROSSEYE tracking nulls, in the absence of any target return, lie on the peaks of the antenna's Δ pattern.[9] The null occurs there because the two sources, E_1 and E_2, of opposite phase (Figure 4.18), separated by the very small angle 2δ, cancel each other when they are symmetrically disposed about a Δ pattern peak. As soon as the target return component E_s is added, the cancellation is upset and can only be restored by deflecting the antenna to move the three sources off the Δ pattern peak (Figure 4.19).

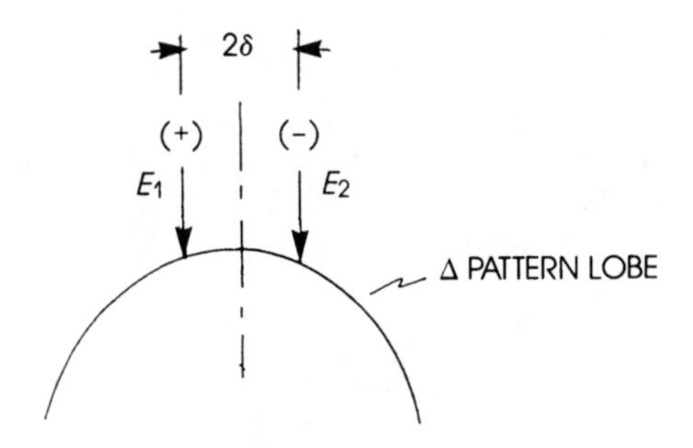

Figure 4.18 Δ-Null condition for pure CROSSEYE.

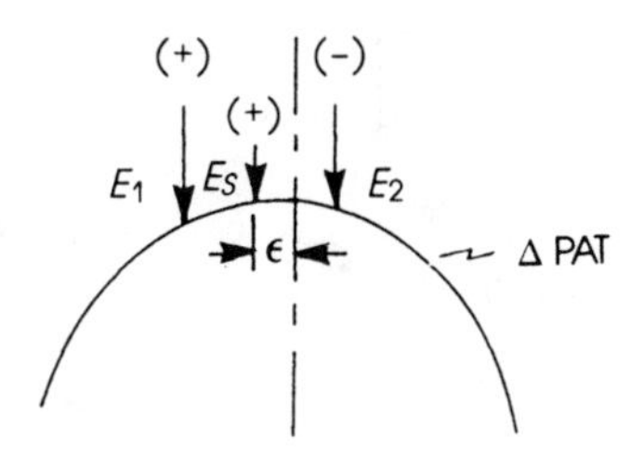

Figure 4.19 Δ-Null condition for finite J/S.

If we let $f_\Delta(\theta)$ represent the Δ pattern sidelobe peak and ϵ be the deflection required to restore cancellation, the cancellation condition is described by

$$E_1 f_\Delta(\delta + \epsilon) + E_2 f_\Delta(\delta - \epsilon) + E_s f_\Delta(\epsilon) = 0$$

The Δ pattern in the vicinity of the peak can be described by the quadratic:

[9]The first peak is the hump of the Δ pattern S curve. Other peaks are sidelobe peaks.

$$f_\Delta(\theta) = 1 - 1.17\left(\frac{\theta}{\beta_\Delta}\right)^2$$

where β_Δ is the 3-dB width of the lobe. With E_2 set equal to $-E_1$ and $E_s f_\Delta(\epsilon)$ approximated by the constant E_s, the balance condition becomes

$$\frac{E_1}{E_s} = \sqrt{J/S} = \frac{1}{4 \times 1.17 \times \delta/\beta_\Delta \times \epsilon/\beta_\Delta}$$

The f_Δ lobe width β_Δ can be shown[10] to be only about 0.65β, so a balance point that lies 90% of the way out to the first CROSSEYE tracking null (measured from the origin, or $\alpha = 0$) lies at a distance:

$$\epsilon = 0.1\alpha_1 = 0.1(0.62\beta) = 0.1\left(\frac{0.62}{0.65}\beta_\Delta\right) = 0.095\beta_\Delta$$

Therefore, the computed J/S needed to get 90% of the way to the first CROSSEYE tracking null is

$$J/S = \left[\frac{1}{4 \times 1.17 \times \delta/\beta_\Delta \times \epsilon/\beta_\Delta}\right]^2 = \left(2.25\frac{\beta_\Delta}{\delta}\right)^2$$

The value of β_Δ/δ in the simulation discussed earlier was approximately 480. This results in a computed J/S requirement of

$$J/S \approx 61 \text{ dB}$$

which is within a few dB of the value obtained from Figure 4.17, based on the simulation. Lower J/S can be effective, but as J/S is decreased the potential tracking error achievable also decreases.

When is the CROSSEYE error large enough? One answer might be that the error is large enough if, after having achieved that error, CROSSEYE could be turned off, leaving the radar to lock a sidelobe tracking null to the target; this is decidedly a break-lock situation. For the tracker to pull from a resting state to the first sidelobe null, the resting point must be near enough to the sidelobe null to ensure that the error signal has the proper polarity to pull toward the sidelobe null and not back toward the primary (boresight) null. In our simulation model, the

[10]Lobe widths quoted here are based on the $(\sin x)/x$ elemental beams used in the simulation.

error signal switches polarity at $\alpha = 1.24\beta$. If released with the antenna resting at a slightly larger off-boresight angle than this, the tracker will pull into the sidelobe lock. Even with infinite J/S, CROSSEYE can only attain a steady-state pull-off of $\alpha_1 = 0.62\beta$. It is not reasonable to expect that application of a CROSSEYE step input would cause the 100% overshoot needed to reach the critical point, $\alpha = 1.24\beta$. Nevertheless, the simulation model indicates that the potential tracking errors attainable by CROSSEYE are of a magnitude that could disable a radar-based fire control system. A model with an antenna pattern different from the (sin $x)/x$ pattern used for the simulation might well come to a different conclusion as to the likelihood of creating break locks.

We mentioned earlier that high J/S requirements constitute just one of the difficulties of implementing CROSSEYE. The other difficulties have to do with the need for close tolerances in amplitude and phase matching of the two emitters. To gain some understanding of these requirements, let us return to the model used for the simulation, in which the target return is depicted as coming from a point scatterer midway between the two CROSSEYE emitters. We saw that the pure CROSSEYE illumination is an odd function of distance from the center of the radar antenna aperture, and the target return produces an even function.

It is easy to show that either a small amplitude unbalance or a small phase mismatch in the CROSSEYE system will upset the illumination of the radar aperture by the addition of an even component of illumination—and the greater the mismatch, the greater the even component. This even component can be viewed as a fictitious target component against which CROSSEYE must compete.

Of course, increasing CROSSEYE power does not help, for the fictitious target would also increase, maintaining a fixed ratio to the ideal (odd-symmetry) CROSSEYE component. We can increase the fictitious $(J/S)_u$ ratio due to unbalance only by controlling the magnitude of the unbalance. In the simulation model, E_1 was the field of one emitter at the center of the radar antenna aperture, and it was assumed that E_s, the target return field, was in phase with E_1. The CROSSEYE field (the combination of E_1 and E_2) was written as

$$E_{\text{odd}} = E_0 \pi \frac{dx}{\lambda R}$$

where $E_0 = 2E_1$ is the field on the peaks of the CROSSEYE lobe pattern. Although we did not say so earlier, this field is in time quadrature to field E_1, so it is in time quadrature to E_s. If we designate the target return field as

$$E_{\text{even}} = E_s$$

then $E_{\text{odd}}/E_{\text{even}}$ is related to J/S by

$$\frac{E_{\text{odd}}}{E_{\text{even}}} = 2E_1\pi\frac{dx}{\lambda R} \div E_s = 2\pi\frac{dx}{\lambda R}\sqrt{J/S} \tag{4.15}$$

We shall be comparing the E_{odd}-to-E_{even} ratios resulting from unbalance against this expression.

Amplitude Mismatch

Let us assume perfect phase matching but a finite amplitude mismatch such that the amplitude ratio is $1 + \delta$, where $\delta \ll 1.0$. The far-field phasor addition at the radar antenna aperture can be written as

$$E = E_1\left(1 + \frac{\delta}{2}\right)\exp\left(j\frac{\pi d\phi}{\lambda}\right) - E_1\left(1 - \frac{\delta}{2}\right)\exp\left(-j\frac{\pi d\phi}{\lambda}\right)$$

By setting $\phi = x/R$, we approximate this sum as

$$E = j2E_1\frac{\pi dx}{\lambda R} + E_1\delta\cos\left(\frac{\pi dx}{\lambda R}\right)$$

Note the quadrature time-phase difference (the $j = \sqrt{-1}$ factor associated with the first term).

We can approximate the even (cosine) term in the region of interest as a constant:

$$E_{\text{even}} = E_1\delta$$

The magnitude of the odd term is

$$E_{\text{odd}} = 2E_1\pi\frac{dx}{\lambda R}$$

as in the absence of mismatch. The ratio of odd to even components is (compare with (4.15)):

$$\frac{E_{\text{odd}}}{E_{\text{even}}} = 2\pi\frac{dx}{\lambda R}\left(\frac{1}{\delta}\right) = 2\pi\frac{dx}{\lambda R}\sqrt{(J/S)_u}$$

where now the fictitious $(J/S)_u = 1/\delta^2$ is due to the amplitude unbalance δ. The effect of a given $(J/S)_u$ is identical to the effect of that value of J/S in the simulation. To attain the equivalent of a 45-dB J/S requires that

$$20 \log \left(\frac{1}{\delta} \right) = 45 \text{ dB}$$

or

$$\delta = 5.62 \times 10^{-3}$$

That is, the amplitudes are balanced to

$$20 \log (1 + \delta) = 0.05 \text{ dB}$$

It is difficult to imagine maintaining such fine balance.

Phase Mismatch

Now assume perfect amplitude balance but a phase mismatch, ψ. The resulting field can be expressed as

$$E = E_1 \exp \left[j \left(\frac{\pi d}{\lambda} \sin(\phi) + \frac{\psi}{2} \right) \right] - E_1 \exp \left[j \left(\frac{\pi d}{\lambda} \sin(\phi) - \frac{\psi}{2} \right) \right]$$

After replacing $\sin(\phi)$ by its argument, we have

$$E = j2E_1 \sin \left(\frac{\pi d}{\lambda} \phi + \frac{\psi}{2} \right)$$

The value of $\phi = x/R$ in the region of interest is small and so is ψ, so the magnitude of E is, approximately,

$$|E| = 2E_1 \left[\frac{\pi dx}{\lambda R} + \frac{\psi}{2} \right]$$

The odd and even components are

$$E_{\text{odd}} = 2E\pi \frac{dx}{\lambda R} \qquad \text{and} \qquad E_{\text{even}} = E_1\psi$$

Their ratio is (compare to (4.15)):

$$\frac{E_{\text{odd}}}{E_{\text{even}}} = 2\pi \frac{dx}{\lambda R}\left(\frac{1}{\psi}\right) = 2\pi \frac{dx}{\lambda R} \sqrt{(J/S)_u}$$

where $(J/S)_u = (1/\psi)^2$.

Now the fictitious $(J/S)_u$ is due to the phase mismatch. This situation is somewhat different from the amplitude unbalance case, for now the even and odd field components are in time phase. If the requirement on $(J/S)_u$ is set at 45 dB, we have

$$20 \log\left(\frac{1}{\psi}\right) = 45$$

or

$$\psi = 10^{-2.25} \text{ rad} \approx 0.3°$$

Again, maintaining this fine tolerance on phase matching is difficult to imagine. The 45-dB value of $(J/S)_u$ was chosen because the simulation indicated that this value is required in order to produce monopulse pointing errors just half as large as are possible with ideal CROSSEYE.

4.6.2 Cross-Polarization (CROSS-POL) Jamming

An electromagnetic traveling wave is characterized by its intensity, its direction of propagation, and its polarization. Its energy is equally shared between its electric and magnetic fields. The electric and magnetic field vectors vibrate in time phase. They are spatially orthogonal to each other, both lying in the plane of the wavefront, which is normal to the direction of propagation. The polarization state of the wave is defined by the motion of the electric field vector (E-vector) in any wavefront plane along the propagation path. If the E-vector vibrates along a line, the wave is said to be linearly polarized. The linear polarization may be horizontal (H), vertical (V), or any other orientation. If the E-vector traces out a circle, the polarization is said to be circular, right-hand (R), or left-hand (L), depending on the direction of rotation of the E-vector. In general, the E-vector traces out an ellipse. Linear and circular polarization are special cases of elliptical polarization. Elliptical polarization is characterized by the axial ratio, the tilt of the axes of the polarization ellipse, and the direction of rotation of the E-vector.

Antenna polarization designates the polarization of the wave that the antenna is designed to transmit or receive. The most common polarizations are H and V linear. Less common are R and L circular. H and V are one example of an orthogonal polarization pair. R and L form another orthogonal pair. Any desired elliptical polarization can, in principle, be synthesized from a combination of components of an orthogonal pair, with appropriate weighting of relative amplitude and relative phase. If one makes the amplitude and phase weights adjustable, the resultant polarization can, in principle, be modulated through all possible polarization states. In practice, this process would be limited by the attainable polarization purity of the two orthogonal components.

Perfect purity of design polarization is an unattainable ideal. The antenna always exhibits a finite response to the undesired orthogonal polarization. The desired response is called the antenna's *design polarization, principal polarization,* or *co-polarization response.* The undesired orthogonal polarization response is called the *cross-polarization response.* The antenna's polarization purity is the ratio of its principal polarization response to its cross-polarization response, sometimes expressed in dB. In linear polarization, the purity is usually expressed as an axial ratio (ratio of H and V axis components).

Orthogonal polarization (CROSS-POL) jamming depends for its effect on the fact that the antenna pattern of the victim radar for orthogonal polarization is quite different from the principal polarization pattern. Van Brunt [2] describes CROSS-POL tactics that may be exercised against search radar functions. We shall be concerned here with CROSS-POL effects against tracking radars, particularly, monopulse tracking radars.

We briefly described monopulse angle tracking in a single plane in Section 4.6 in terms of sum (Σ) and difference (Δ) processing, where the Σ and Δ channel responses are derived from a pair of identical beams, squinted on either side of the boresight axis. The Σ channel response peaks on the boresight axis, and the Δ channel has a null on that axis. Tracking capability depends on these well-defined patterns that prevail for the principal polarization. The radar transmits via the Σ channel, illuminating the target with the principal polarization. Backscatter from the target may be predominantly at the incident polarization, but most targets also yield a signficant orthogonal backscatter component. The orthogonal component, entering the radar antenna via its finite orthogonal polarization response, does induce a finite tracking error, but the error is small if the antenna exhibits a high degree of polarization discrimination.

An airborne self-protection CROSS-POL jammer aims to inject into the victim monopulse receiver an orthogonal polarization component as large as or larger than the co-polarized component of the target return. To do so, the jammer must overcome the radar antenna's cross-polarization discrimination capability (its polarization purity ratio). Moreover, the jammer must, as far as possible, refrain from transmitting co-polarized energy, for this energy acts as a beacon emission. Thus, polarization purity is as important to the jammer as to the radar. The jammer would normally assume the radar's polarization to be identical to the polarization of the radar emission received by the jammer. The jammer, acting in a repeater mode, can transform the received radar polarization to the orthogonal jamming polarization by the scheme depicted in Figure 4.20. Receiving antennas

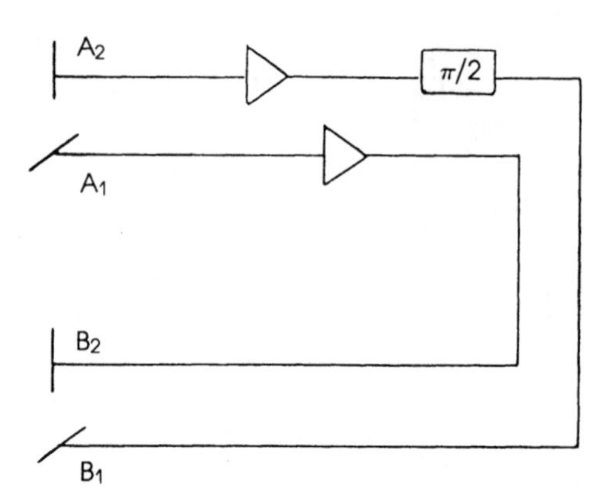

Figure 4.20 CROSS-POL repeater jammer.

A_1 and A_2 are an orthogonal pair, as are transmitting antennas B_1 and B_2. In this case the orthogonal pairs are H and V linear. The H component received on A_1 is amplified and transmitted as a V component by B_2, and the V component received on A_2 is amplified and transmitted as an H component by B_1. The gains through the two paths are matched, as are the phase lags, except for an added 180° lag inserted in one path. There are obvious difficulties in implementing such a CROSS-POL repeater scheme. The jammer need not operate in a repeater mode. We can use any jamming waveform, including continuous noise, for CROSS-POL jamming, provided that we maintain the jammer polarization orthogonal to the radar polarization.

We can understand the ability of CROSS-POL jamming to create monopulse tracking errors by comparing the co- and cross-polarization responses of antennas. For a paraboloidal reflector antenna, the cross-polarized response arises from polarization impurity of the primary feed and from polarization conversion by the reflector surface. Ghobrial [3] presents computed plots of the co-polar and cross-polar patterns of a perfect paraboloidal reflector[11] illuminated by a dipole feed. The dominant features of the cross-polar pattern are a set of four lobes spaced about the main lobe of the co-polar pattern, with their peaks lying on lines oriented at ±45° relative to the principal (H and V) axes of the co-polarized main lobe. These cross-polarized lobes are sometimes called *Condon lobes.* The cross-polar response vanishes along the principal axes.

Bodnar [4] explains that the cross-polar lobes alternate in polarity (adjacent lobe responses exhibit 180° relative phase shift). He presents a computed plot of

[11]Imperfections in the reflector surface can produce additional cross-polar components.

the cross-polar monopulse azimuth difference pattern. The general shape of this six-lobed pattern can be deduced from the combination of the sets of four-lobed cross-polarization patterns of the pair of squinted beams that form the monopulse difference channel. In another paper [5] Bodnar presents measured cross-polar patterns for the sum channel and for azimuth and elevation difference channels of a monopulse antenna. The measured patterns have some of the lobe features predicted by computation, but they lack the perfect symmetry of the computed patterns.

Figure 4.21 shows the locations and general shapes of the co-polar (solid contours) and cross-polar (dashed contours) lobes of monopulse sum and difference patterns, based on idealized pattern computations. The + and − signs indicate the relative phases of the various lobes. Clearly, if a jammer were able to inject pure cross-polarized energy so strong as to overshadow the co-polar target return, the monopulse tracking axis would come to rest at an intersection of the stable null loci of the azimuth and elevation cross-polar difference patterns.

As we examine Figure 4.21, we can only be certain that the cross-polar Δ channel nulls lie in the valleys between the Δ channel lobes. Whether a null locus represents stable tracking depends on the Σ channel polarity in the region of the Δ channel null locus. The information in Figure 4.21 is too scanty to enable us to sketch the loci of the cross-polar tracking nulls, but it is likely that the intersections of az-el tracking loci would fall well off the co-polar boresight axis, perhaps as much as a beamwidth away, among the valleys of the cross-polar Δ channel lobe structure. Hence, the expectation that powerful CROSS-POL jamming would produce large monopulse tracking errors and even break lock. Van Brunt [2] asserts that it is very difficult to produce break lock. Not only is it difficult to produce a high degree of polarization purity, but it is also difficult to achieve high gain in the repeater jammer case without causing instability (because of transmitting-receiving antenna coupling).

In a laboratory environment, it would be possible, in principle, to transmit a jamming polarization with just enough co-polar component to cancel the co-polar target return (and perhaps clutter return) and leave a residue of pure cross-polarized energy within the radar's range gate. In a field environment this condition is impossible to maintain, but it may be approached momentarily if the jamming polarization is modulated, as suggested earlier, through all possible polarization states. With a jog detector to reveal the polarizations that produce large deflections of the radar beam, the extent of the polarization modulation can be restricted to a small range of productive polarization states. If the radar's transmitting and receiving polarizations are the same, the incident radar polarization observed by the jammer gives a good indication of the orthogonal polarization that the jammer should transmit. This clue has no value if the radar has a polarization diversity capability when receiving. In fact, if the radar receiver were to have a receive polarization agility, even the jog detector would lose its value. In a bistatic radar system, the transmitted

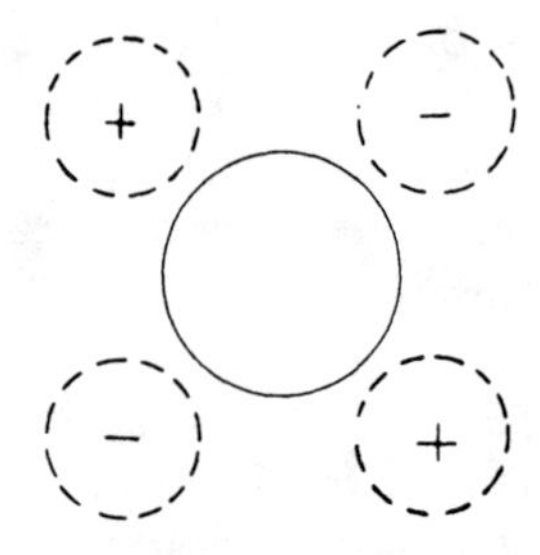

(a) Sum Pattern

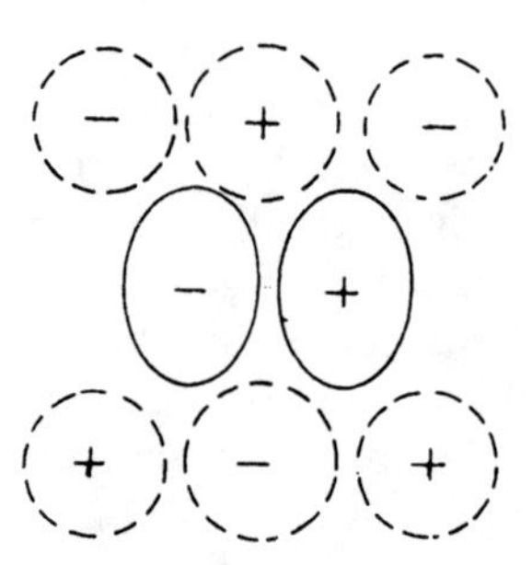

(b) Azimuth Difference Pattern

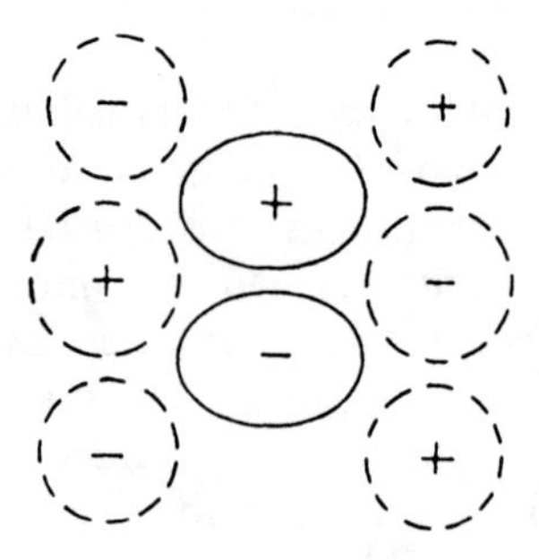

(c) Elevation Difference Pattern

Figure 4.21 Monopulse beam patterns.

polarization cannot be taken as an indication of received polarization. For instance, in a semiactive radar-guided missile system, even if the missile seeker antenna has the same nominal polarization as the target illuminator, missile attitude variation causes the receiver polarization to vary. If the target aircraft employs CROSS-POL jamming, a jog detector might help the jammer to deflect the beam of the target illuminator, but as long as an adequate level of illumination exists the semiactive seeker of the missile continues to function. In fact, the missile may successfully home on the jamming emission.

Let us consider an airborne CROSS-POL jammer that produces cross-polarized power density J_x at the victim radar antenna, along with an inadvertent co-polarized component J_0, and let the co-polarized target return power density be S. The jammer's polarization purity ratio is

$$\rho_J = \frac{J_x}{J_0}$$

If $\rho_J \gg 1.0$, J/S at the input to the radar antenna is

$$J/S \approx \frac{J_x}{S}$$

Both J_0 and S are passed by the antenna to the receiver with unity weight, but the component J_x is weighted by $1/\rho_R$, where ρ_R is the radar antenna polarization purity ratio. For large tracking errors to be produced, the cross-polarized component reaching the angle discriminator should exceed both the co-polarized signal and the jamming components. This requirement leads to the inequalities:

$$J_x \cdot \frac{1}{\rho_R} > S \quad \text{and} \quad J_x \cdot \frac{1}{\rho_R} > J_0$$

The first inequality calls for J/S greater than the polarization purity ratio of the radar antenna:

$$J/S \approx \frac{J_x}{S} > \rho_R$$

The second inequality calls for a jammer antenna with a polarization purity ratio better than that of the radar antenna:

$$\rho_J = \frac{J_x}{J_0} > \rho_R$$

Just how large J/S and ρ_J should be compared with ρ_R depends on the magnitude of the tracking errors one hopes to create. In any event, the better the polarization purity of the radar antenna, the more difficult is the task of the CROSS-POL jammer.

Polarization purity of an antenna is sometimes specified only for the main-lobe region. Ghobrial [3] shows that, for an ideal, symmetrical, reflector antenna, cross-polarization nulls lie along the principal (H and V) axes of the main lobe for H and V polarizations. For circular polarization the patterns have circular symmetry, with a cross-polarization null on the boresight axis. Thus, good polarization purity is expected in the central portion of the main lobe. In the main lobe regions, paraboloidal reflector antennas may have purity ratios of 20 to 30 dB. Flat-plate array antennas may have purity ratios of 30 to 50 dB. Ghobrial [3], Bodnar [4], and others have shown that for paraboloidal reflectors polarization purity is improved by increasing the focal-length-to-diameter ratio (F/D). Note that even with a target (or jammer) on boresight (on the tracking axis), energy is not received on the centers of the squinted lobes that form the Σ and Δ patterns of a monopulse tracker.

Figure 4.22 shows cross-polar responses for three elevation cuts (planes) on boresight and at $\pm 3°$ relative to boresight for a specific radar antenna. In this case the design polarization is linear (vertical), so polarization purity can be stated as an

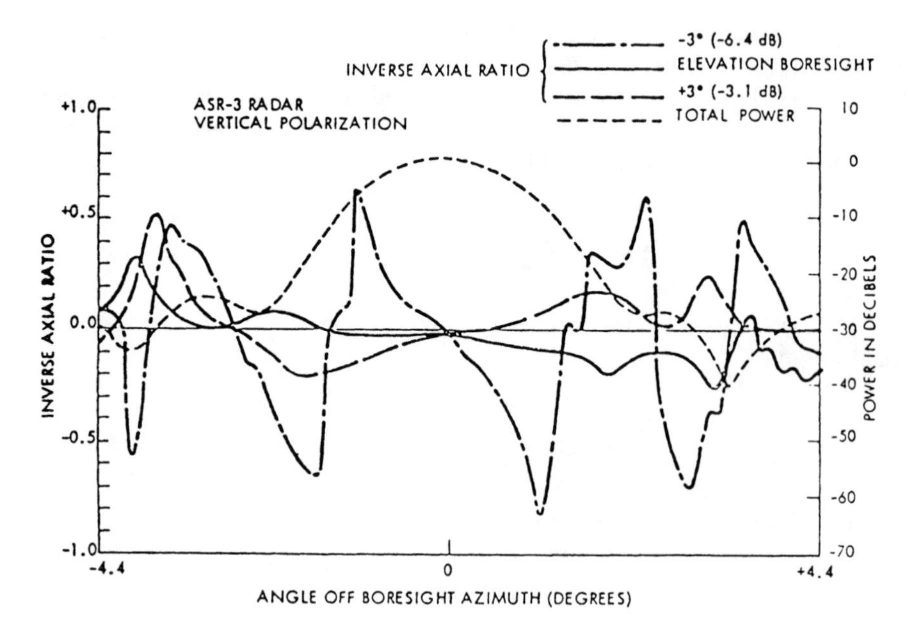

Figure 4.22 Cross-polarization response of ASR-3 radar antenna. (Courtesy of Advent Systems, Inc.)

axial (V/H) ratio. In the plot, the inverse (H/V) ratio is used as the measure of polarization impurity. Note the cross-polarization null on azimuth boresight for all three elevation cuts. The polarization purity within the 3-dB azimuth beamwidth is about 20 dB for the elevation boresight cut. In the sidelobe region the purity degrades to about 10 dB for that cut. The $-3°$ elevation cut contains cross-polarization peaks that are only 2 to 6 dB below the co-polarized main-lobe peak.

Although much of the foregoing discussion deals with the effect of CROSS-POL against monopulse, it should be evident that it will also produce angle errors in other forms of tracking radars. For instance, a CONSCAN radar might come to equilibrium with its boresight (scan) axis pointing far enough off target to cause one of its cross-polar lobes to scan about the target's CROSS-POL jammer.

4.7 CONOPULSE

A major virtue of CONSCAN is its ability to carry all target angle information in a single receiver channel. Its major disadvantage is its susceptibility to amplitude-induced measurement errors. Monopulse overcomes the effects of signal amplitude fluctuation at a cost of a considerably more complex antenna feed structure and, in its most straightforward form,[12] three separate receiver channels (Σ, Δ_{az}, and Δ_{el}).

Figure 4.23 represents a conical scan radar that employs a pair of squinted beams rather than a single beam.[13] This arrangement has been called [6] *conopulse,* because it embodies features of both CONSCAN and monopulse. In Soviet literature it has been called *scan with compensation.* The conopulse system is analyzed in Ref. 6 for two forms of signal processing. Conopulse requires two receiver channels, which may be the v_{s1}, v_{s2} channels of beams 1 and 2 of Figure 4.23, or Δ and Σ channels formed from these beams.

The two beams are symmetrically squinted on either side of the scan axis at angle θ_s. If the beam pair were stopped with their axes in the horizontal plane, they could serve as a monopulse pair for measuring target azimuth, or if oriented in the vertical plane they could provide an elevation measurement. The continuous rotation of the beams at angular scan frequency ω_s allows them to be shared between the azimuth and elevation measurement functions. As with CONSCAN, a complete rotation produces a scan modulation cycle from which both azimuth and elevation tracking error components can be determined (in practice, the measurement is averaged over many scan cycles).

The major virtue of conopulse, compared with CONSCAN, is its immunity to amplitude-induced tracking errors. As with monopulse, neither target amplitude

[12]Multiplexing schemes have been devised to permit the two Δ signals to share a common receiver channel, resulting in two-channel monopulse.

[13]Compare Figures 4.23 and 4.3.

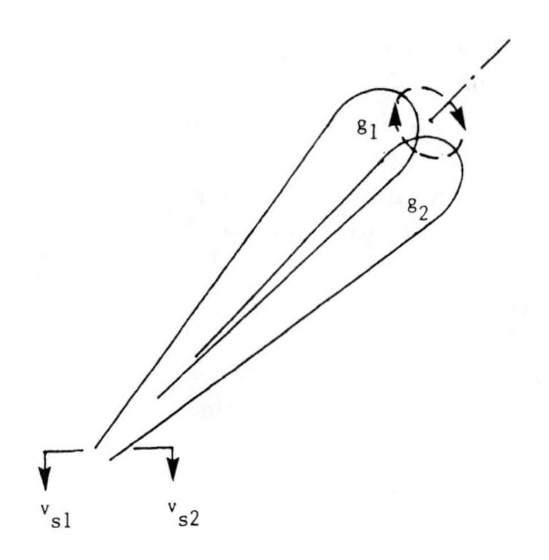

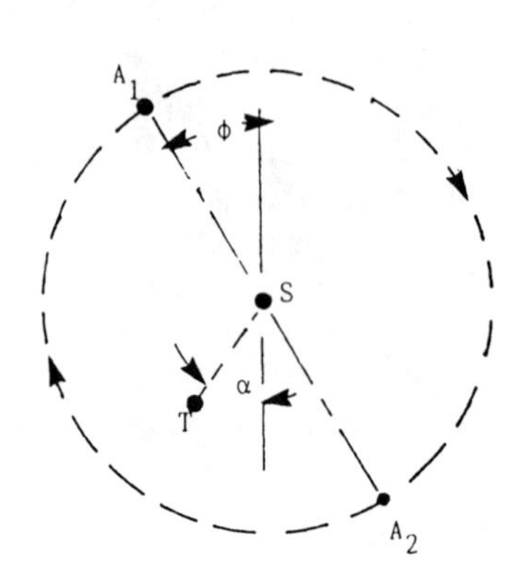

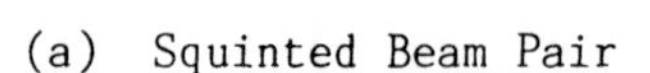

(b) Target Plane

Figure 4.23 Conopulse geometry.

fluctuation nor AM jamming originating at the target will generate angle errors in the ideal conopulse system. We can demonstrate this immunity by referring to the small-error model used in Section 4.2 when we discussed CONSCAN. Let θ_1 be the angle of the target relative to the axis of beam 1. The value of θ_1 as a function of scan angle $\phi = \omega_s t$ is identical to the value given in (4.3):

$$\theta_1^2 = \theta_T^2 + \theta_s^2 + 2\theta_T\theta_s \cos(\omega_s t + \alpha)$$

The angle θ_2 of the target relative to the axis of beam 2 is then given by

$$\theta_2^2 = \theta_T^2 + \theta_s^2 - 2\theta_T\theta_s \cos(\omega_s t + \alpha)$$

We shall assume that the target is illuminated by a nonscanning transmitting beam (as in COSRO). The signal from beam 1 is

$$v_{s1} = v_0 g_1(\theta_1)$$

where v_0 is the target return amplitude that would exist if the target were on the peak of beam 1, and $g_1(\theta)$ is the normalized voltage pattern of the beam (the beam has axial symmetry). Likewise, for beam 2:

$$v_{s2} = v_0 g_2(\theta_2)$$

Let us assume that the beam patterns g_1 and g_2 are identical:

$$g_1(\theta) = g_2(\theta) = g(\theta)$$

At some point in the signal processing (it can be at the antenna or further down line), sum and difference channels are formed:

$$\Delta = v_{s2} - v_{s1} = v_0[g(\theta_2) - g(\theta_1)]$$
$$\Sigma = v_{s2} + v_{s1} = v_0[g(\theta_2) + g(\theta_1)]$$

Clearly, when we normalize Δ by dividing by Σ, the target return amplitude is removed from consideration, so neither target amplitude fluctuation nor AM jamming from the target vehicle will influence the ratio Δ/Σ, which is used to develop the azimuth and elevation error signals.

Let the beam pattern be approximated by

$$g(\theta) = e^{-a\theta^2}$$

where

$$a = \frac{2\ln(2)}{\beta^2}$$

and β is the 3-dB beamwidth.[14] The quadratic approximation

$$g(\theta) = 1 - a\theta^2$$

is valid in the central portion of the main lobe, so the difference pattern modulation becomes

$$\Delta = v_0[a\theta_1^2 - a\theta_2^2] = v_0 a 4\theta_T\theta_s \cos(\omega_s t + \alpha)$$

the sum channel output becomes

$$\Sigma = v_0[2 - a(\theta_1^2 + \theta_2^2)] = v_0 2[1 - a(\theta_T^2 + \theta_s^2)]$$

[14]Target modulation is affected by the one-way beam pattern, for the target is illuminated by a separate nonscanning beam.

Thus, Δ/Σ contains the $\cos(\omega_s t + \alpha)$ term from which we can derive the azimuth and elevation error signals exactly as in CONSCAN.

Although the foregoing discussion employs a small-angle error approximation, the immunity of conopulse to amplitude effects does not depend on the tracking error being small. Reference 5 provides an analysis that is not restricted to small-angle errors.

In Section 4.2 we noted that the scanning RF feed should execute a nutating motion rather than a simple rotation. This is a precaution against the generation of polarization-induced AM that is periodic at the scan frequency, but with phase related to target return polarization, not to target angle ϕ. This argument does not apply to conopulse, because, with the two feeds maintaining identical polarizations, any polarization-induced AM would be identical for the two beams; consequently, the Δ/Σ processing would remove the effect of that modulation.

If the conopulse system operates on a COSRO basis, with a nonscanning transmitting beam of fixed polarization, an argument could be made for nutating scanning feeds that have a polarization matching the transmitted polarization. The basis of this argument is that for many targets the polarization of the dominant component of the radar return is the transmitted polarization. If the feed polarization were allowed to rotate, at two points on the scan cycle the received polarization (assuming linear polarization) would be orthogonal to the dominant component of the radar return. Based on the discussion of CROSS-POL jamming (Section 4.6.2), one would expect increased tracking error in the plane for which the polarization mismatch exists. On the other hand, if the transmission is via the Σ combination of the two scanning beams, either nutation or simple rotation would result in the received polarization being matched to the transmitted polarization, in which case there seems to be no argument favoring one form of feed motion over the other.

As for ECM susceptibilities, an ideally implemented conopulse radar should have essentially the same strengths and vulnerabilities as a monopulse radar. It is immune to amplitude-modulated jamming, and its response to CROSSEYE and CROSS-POL jamming should be comparable to the response of a monopulse radar.

REFERENCES

1. Barton, D.K., *Modern Radar System Analysis,* Artech House, Norwood, MA, 1988.
2. Van Brunt, L.B., *Applied ECM,* EW Engineering, Dunn Loring, VA.
3. Ghobrial, S.I., "Co-Polar and Cross-Polar Diffraction Images in the Focal Plane of Paraboloidal Reflectors: A Comparison between Linear and Circular Polarization," *IEEE Transactions on Antennas and Propagation,* Vol. AP-24, No. 4, July 1976, pp. 418–424.

4. Bodnar, D.G., "Cross-Polarized Characteristics of Monopulse Difference Patterns," *IEEE Antennas and Propagation Symposium Record,* 1980, pp. 477–480.
5. Bodnar, D.G., "Polarization Characteristics of Monopulse Tracking Antennas," *Microwave Journal,* Vol. 27, No. 12, December 1984, pp. 123–136.
6. Sakamoto, H., and P.Z. Peebles, Jr., "Conopulse Radar," *IEEE Transactions,* Vol. AES-14, No. 1, January 1978, pp. 199–208.

Chapter 5
VELOCITY DECEPTION

5.1 ORIGIN AND USES OF DOPPLER SHIFT

The signal observed at a receiver exhibits a doppler shift whenever there is a time-varying path length between transmitter and receiver. In Figure 5.1, both the radar and the target are moving, so the length, $2R$, of the propagation path is varying with time. The doppler shift can be attributed to the time-varying phase lag, ϕ, of the propagation path. This lag, in the radar case, is

$$\phi = \frac{2\pi}{\lambda}(\text{path length}) = \frac{2\pi}{\lambda}(2R) = \frac{4\pi}{\lambda}R$$

where λ is the wavelength. The wavelength is related to the angular frequency, ω_0, of the transmitted signal by

$$\lambda = \frac{2\pi c}{\omega_0}$$

where c is the velocity of light. If the transmitted signal is represented by

$$E_X = \cos(\omega_0 t) \tag{5.1}$$

then the received signal can be represented as

$$E_R = A\cos(\omega_0 t - \phi) = A\cos\left(\omega_0 t - \frac{4\pi}{\lambda}R\right) \tag{5.2}$$

For any sinusoidal function of time, the angular frequency is, by definition, the time derivative of the argument. Hence, from (5.1), the transmitted angular frequency is

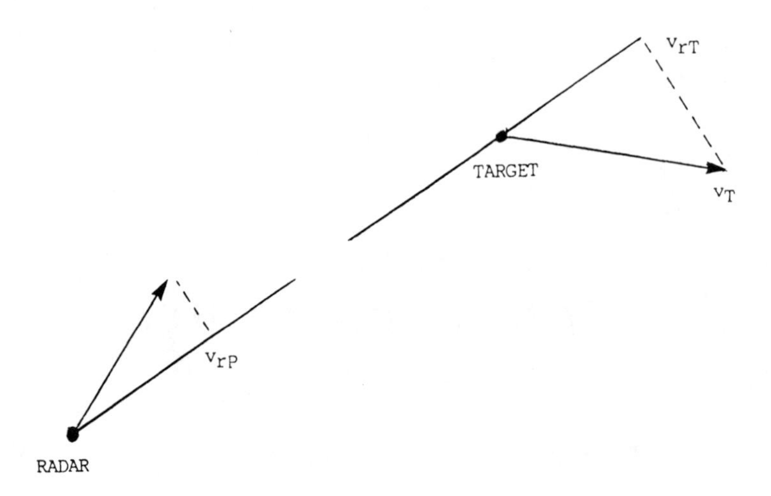

Figure 5.1 Time-varying propagation path length.

$$\omega_X = \frac{\mathrm{d}}{\mathrm{d}t}(\omega_0 t) = \omega_0$$

and the received frequency, from (5.2), is

$$\omega_R = \frac{\mathrm{d}}{\mathrm{d}t}\left(\omega_0 t - \frac{4\pi}{\lambda} R\right) = \omega_0 - \frac{4\pi}{\lambda}\dot{R}$$

where $\dot{R} = \mathrm{d}R/\mathrm{d}t$ is the range rate.

If the range is constant ($\omega_R = \omega_X$) but $\dot{R} \neq 0$, the received echo exhibits a doppler shift, ω_d, defined as

$$\omega_d = \omega_R - \omega_X = -\frac{4\pi}{\lambda}\dot{R} \tag{5.3}$$

The range rate $\dot{R}$ is simply the radial velocity, v_r, of the target relative to the radar, so (5.3) may be written as

$$\omega_d = -\frac{4\pi}{\lambda} v_r$$

or

$$f_d = \frac{\omega_d}{2\pi} = -\frac{2}{\lambda} v_r \tag{5.4}$$

Often the equation for doppler shift is written without the minus sign that appears in the preceding equations. Generally, this means that the equation expresses the magnitude of the shift without regard to its direction. It is often important to be aware of the direction of the shift. The minus sign in the preceding equations is consistent with the fact that a positive v_r indicates an increasing range separation, which in turn produces a negative doppler shift. In Figure 5.1 the radar platform velocity is v_p, and the target velocity is v_T. The relative radial velocity v_r is the difference of the radial (line-of-sight) components of these two velocities:

$$v_r = v_{rT} - v_{rP}$$

Figure 5.2 illustrates a situation in which target doppler plays an important role in detection and tracking. An airborne radar moving at velocity v_p is operating in a look-down mode searching for low-flying aircraft. The radar's resolution volume is a spherical shell of thickness ΔR, equal to the range resolution and shaped in the orthogonal directions by the antenna pattern. In Figure 5.2(a), the low-flying aircraft is in the main-lobe portion of the resolution volume along with a great deal of ground clutter. In addition to the main-lobe ground clutter, considerable side-lobe clutter may be received within the same range resolution cell. In Figure 5.2(b), the radar antenna main-lobe footprint has a width βR, where β is the angular width of the circular pencil beam and R is the range. The resolution shell of thickness ΔR illuminates a swath of width $\Delta R \sec \delta$, where δ is the beam depression angle, so main-lobe clutter originates from the cross-hatched region of area:

$$A_c = \beta R \, \Delta R \sec \delta$$

If, for instance, $\beta = 3°$, $R = 40$ km, $\delta = 15°$, and $\Delta R = 150$ m,[1] the main-lobe clutter area is

$$A_c \approx 3.25 \times 10^5 \text{ m}^2$$

With a backscatter coefficient (clutter RCS per unit of illuminated area) of

$$\sigma° = 10^{-3} \text{ or } -30 \text{ dB}$$

the main-lobe clutter RCS is

[1]Corresponding to a 1-μs simple pulse.

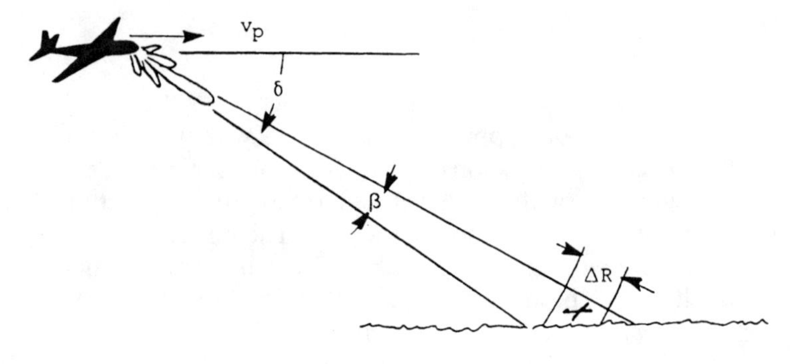

(a) Look-Down Radar Geometry

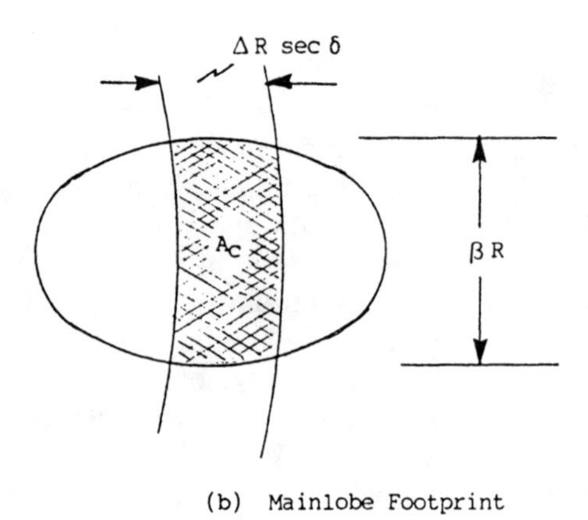

(b) Mainlobe Footprint

Figure 5.2 The look-down detection and track problem.

$$\sigma_c = \sigma^\circ A_c = 325 \text{ m}^2$$

This result is orders of magnitude greater than the RCS of the target aircraft. Target detection and tracking would be out of the question but for the possibility of employing doppler resolution to separate targets from clutter. The radar's mainlobe clutter spectrum will be centered at a positive doppler frequency given by (5.4), with $v_r = -v_{rP}$. The target doppler frequency will be given by (5.4), with

$$v_r = v_{rT} - v_{rP}$$

so it can lie anywhere within a wide range of positive and negative values, depending on the values of v_P and v_T and the orientations of the two velocity vectors. With a narrow main lobe, the main-lobe clutter spectrum is narrow, so there is a low probability that the target doppler will coincide with the doppler of the main-lobe clutter. For surface targets (trucks and tanks) the spread of target velocity is much smaller, so there is a much higher probability that the target doppler will fall within, or near the edges of, the main-lobe clutter spectrum.

The earliest method of dealing with the clutter problem was to employ *moving target indication* (MTI). This is a clutter rejection technique [1, Chapter 17] that places a filter rejection notch at the center of the doppler spectrum of the clutter. The first MTIs were analog devices capable of processing the returns (range sweeps) from only two or three consecutive pulses. Modern MTIs are implemented digitally. They often process larger numbers of pulses, thereby permitting more nearly optimal shaping of the rejection filter characteristic; it is important that the filter not unduly attenuate target returns that have dopplers falling close to the edges of the clutter spectrum.

There is a helpful analogy between linear (one-dimensional) antenna array synthesis and doppler filter synthesis for a pulse radar. A two-element array and a two-pulse MTI form similar interferometer patterns, the former in angle space and the latter in the doppler frequency domain. There are multiple ambiguities. As more and more elements are included in the array, a well-defined beam pattern can be synthesized, with high gain at a specific angle. In the pulsed radar case, as more and more pulses are coherently processed, a well-defined doppler filter can be synthesized with high integration gain at a specific doppler frequency. In both the array and the doppler filter, ambiguities (extra lobes) exist, but as the element spacing in the array and the pulse spacing in the radar become closer and closer, the ambiguities spread further and further apart (in angle/time), perhaps eventually moving out of the region in which target returns are expected. Ultimately, the array approaches a continuous aperture of some width d. The angular resolution (beamwidth) is inversely proportional to d. The limit in the radar signal case is represented by the CW radar that integrates for some observation time T_i. Its doppler resolution is inversely proportional to T_i. In both cases the ambiguous response lobes will have disappeared, although minor lobes (sidelobe responses) will remain on either side of the main lobe.

Concerning coherence requirements for doppler measurement, a coherent CW radar transmits a carrier that can be represented as a true constant-frequency sinusoid. A coherent pulsed radar transmits pulsed samples of such a sinusoid. Random modulation of the phase of the sinusoid, even at low level, degrades coherence. It spreads the line spectrum and degrades doppler resolution. Noncoherent radars (e.g., radars employing magnetron transmitters) transmit pulse trains with random pulse-to-pulse starting phase. This random phase makes doppler processing impossible unless it is measured and removed by a scheme sometimes called *coherent-on-receive processing*.

5.2 CW RADAR VELOCITY DECEPTION

A true CW radar (such as a police radar) transmits an unmodulated carrier and therefore has no range measurement capability. Such a radar depends entirely on target doppler to discriminate against clutter that it receives from all ranges illuminated by the antenna. A military application of CW radar is in missile guidance. Figure 5.3 illustrates a SAM homing on the CW energy reflected off attack aircraft A. Because the missile seeker has no radar transmitter aboard, it is called a *semiactive* seeker; it depends on the transmitter (CW illuminator) at the launch site to illuminate the target. An aft-looking antenna on the missile intercepts a sample of the CW illumination to use as a reference for demodulating the target doppler. Figure 5.4 is a simplified block diagram of such a semiactive seeker.

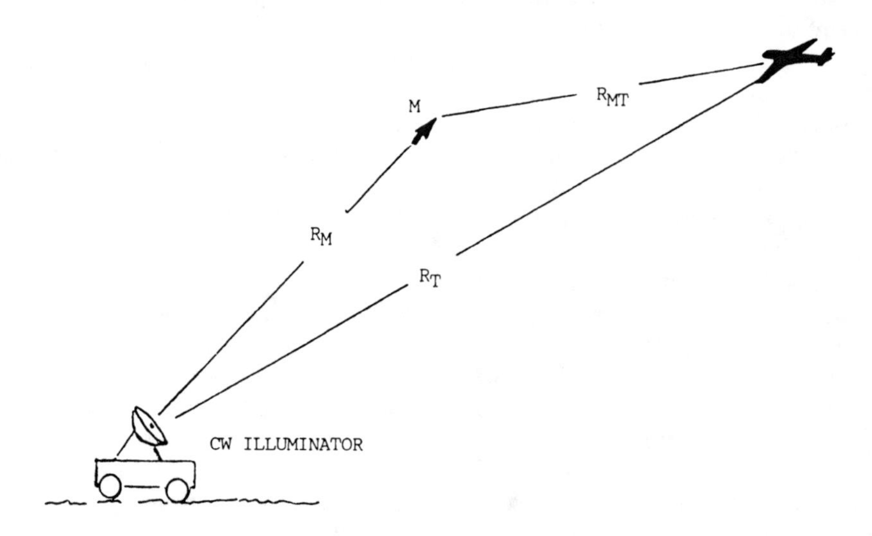

Figure 5.3 Geometry of semiactive radar guidance.

If the illuminator transmits a frequency f_0, the forward-looking antenna of the seeker receives a reflected signal at $f_0 + f_A$, where f_A is proportional to the rate of change of the length of propagation path, $R_T + R_{MT}$, of Figure 5.3. The aft antenna receives a frequency $f_0 + f_B$, where f_B is proportional to the rate of change of the length of path R_M.[2] Both receiving channels are heterodyned by a common local oscillator down to an IF centered at f_i. An *automatic frequency control* (AFC) loop in each path locks to the doppler-shifted CW, $(f_i + f_A)$ or $(f_i + f_B)$, serving as a

[2]The value of f_B will be negative; the missile is receding from the illuminator.

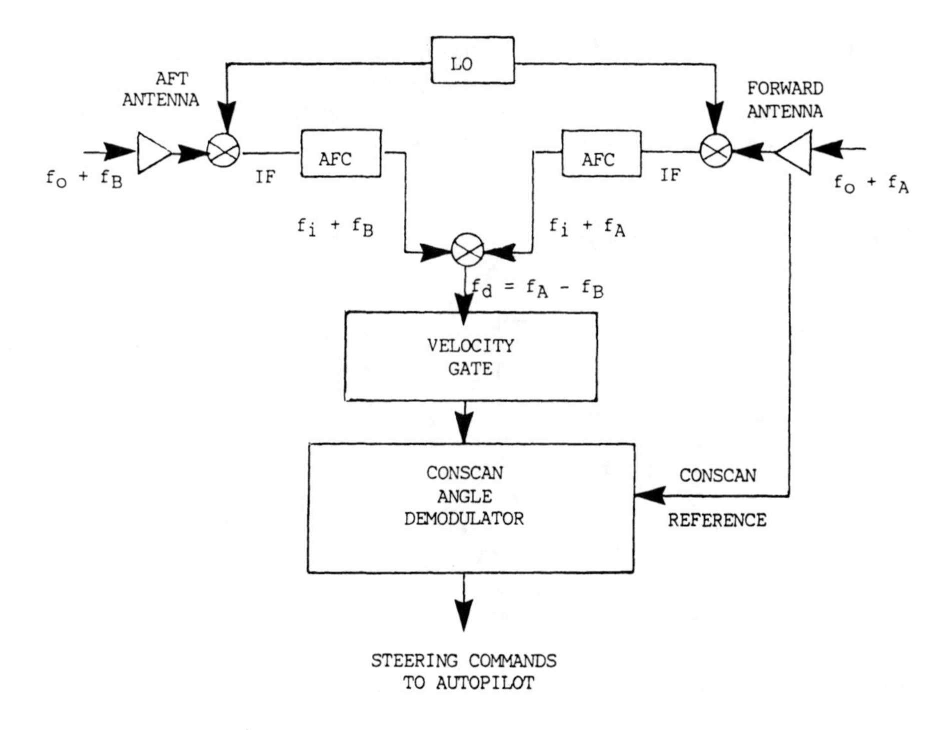

Figure 5.4 Semiactive CW seeker block diagram.

narrow-band tracking doppler filter. A synchronous demodulator, using the $f_i + f_B$ component as a reference, develops the difference doppler output at frequency $f_A - f_B$. This may be a few kilohertz to a few hundred kHz. It bears an amplitude modulation at perhaps 20 to 60 Hz, generated by the conical scan of the forward antenna. This AM is envelope-detected and further demodulated to derive orthogonal angle error signals that serve as steering commands to the missile autopilot. The velocity gate block in Figure 5.4 is a tracking doppler filter that may be implemented as an AFC loop or a *phase-locked loop* (PLL).

A jammer aboard the aircraft may succeed in capturing the velocity gate and pulling it off the target (the skin return from the aircraft). To accomplish this *velocity-gate pull-off* (VGPO), the jammer must initially transmit a false-target CW component at, or near, $f_0 + f_A$ in order to get into the velocity gate. If the false-target signal is sufficiently strong, it will dominate the velocity gate; its frequency can then be slowly shifted, carrying the velocity gate away from the doppler of the target. Figure 5.5 indicates how VGPO might be implemented. Energy received from the CW illuminator is downshifted to a convenient IF and passed through a narrow-band tracking filter, which may be implemented as an AFC loop or a phase-locked loop. The IF carrier from this filter becomes the carrier for a *single-sideband* (SSB) modulator capable of impressing a false-doppler offset on the IF carrier. This IF carrier is then upshifted to RF, using the same LO that was used for the downshift to IF. It is then amplified and transmitted toward the missile.

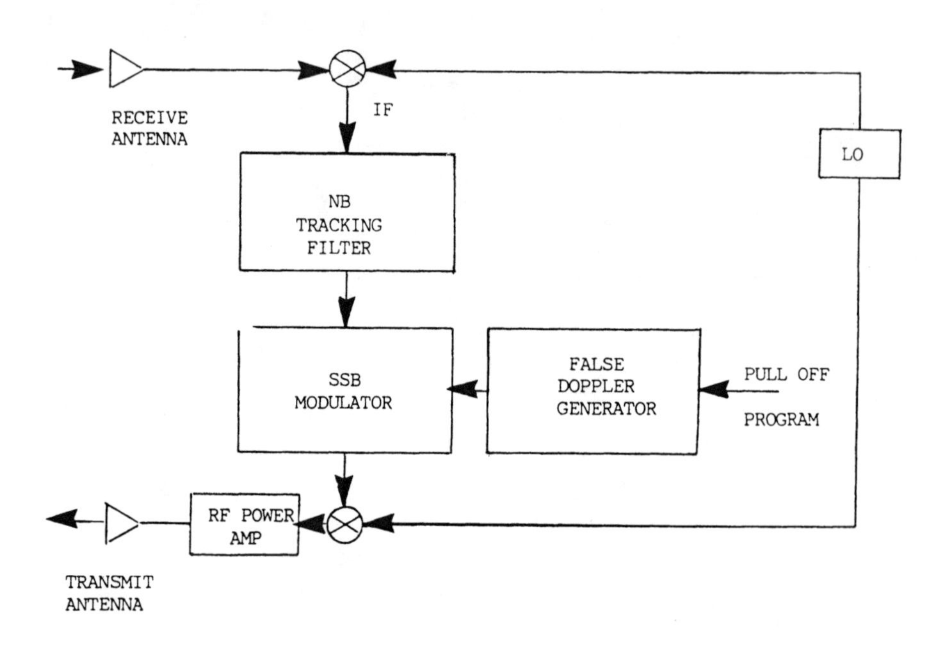

Figure 5.5 CW velocity deception jammer block diagram.

We can also insert the false doppler at RF, by serrodyne modulation of a *traveling wave tube* (TWT) in the RF transmission path. If the false-doppler shift is initially zero, the jammer signal received by the missile is at $f_0 + f_A$, a frequency identical to the skin-return frequency.

The VGPO maneuver by itself would be to no avail if the missile were allowed to continue tracking a clean unmodulated false doppler from the airborne jammer. An unmodulated CW emission from the jammer would serve just as well as the aircraft skin return for angle tracking and would very likely be stronger. One jammer tactic is to pull the missile's velocity gate well off the target doppler and then cease transmitting, leaving the seeker with the task of reacquiring the target. This process, repeated at an appropriate rate, leaves the missile with only intermittent steering information. Another jamming alternative is to apply angle-deception AM to the jammer output, once the end of the VGPO phase is reached. At that point, provided that the VGPO has succeeded, the J/S in the seeker receiver is infinite, so the AM into the seeker's CONSCAN demodulator has a good chance of disrupting the missile's steering commands. Angle deception is covered in Chapter 4.

5.3 PULSED COHERENT RADAR

Although providing excellent doppler resolution and doppler measurement capability, CW radar has two shortcomings that exclude it from many applications. The first is its inability to measure target range or to resolve targets separated in range. In Chapter 3 we pointed out that frequency modulation of the carrier gives the CW radar a ranging capability while maintaining its doppler capability. The second shortcoming is the necessity for transmitting and receiving simultaneously. In most cases, we can achieve sufficient isolation between the transmitting and receiving channels only by using separate transmitting and receiving antennas. This requirement is unacceptable in many applications. Pulsed radar solves the isolation problem neatly by listening (receiving) only when the transmitter is quiet. The pulsed radar sacrifices the privilege of continuous observation of the target return; it can only collect samples at a rate equal to the radar's pulse repetition frequency. As is intuitively clear, if the sampling is frequent enough, little or no target information is lost. However, rapid sampling can lead to ambiguous measurement of target range; echoes from distant targets, elicited by an earlier pulse, may arrive in time coincidence with echoes from nearby targets, elicited by the most recent pulse. To avoid this "multiple time around" problem, we must limit the PRF to

$$\mathrm{PRF}_{\mathrm{max}} \leq \frac{c}{2R_{\mathrm{max}}}$$

where R_{max} is the greatest range from which echoes of significant strength are expected. For instance, if a ground-based, air surveillance radar needs an unambiguous range of 100 nmi, the PRF must be less than 809 pulses per second (809 Hz). This relatively low sampling rate is quite adequate for sampling the low-frequency fluctuations of ground clutter, but, as we shall see, it is too low for unambiguous sampling of the doppler waveform of the return from a high-speed aircraft target.

At X band ($\lambda \approx 0.03$ m), the doppler of a subsonic aircraft ($v_T \leq 330$ m/s) observed by a stationary radar has an upper limit of

$$|f_d|_{\mathrm{max}} = \frac{2v_{T\mathrm{max}}}{\lambda} = 22 \text{ kHz}$$

so the observed doppler may lie anywhere in the range $-22 \text{ kHz} \leq f_d \leq +22$ kHz. Figure 5.6(a) shows a CW radar doppler spectrum containing the carrier at

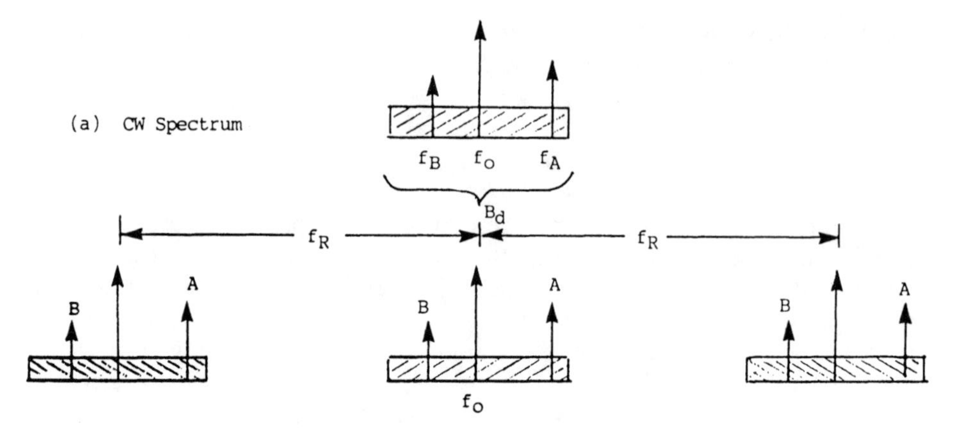

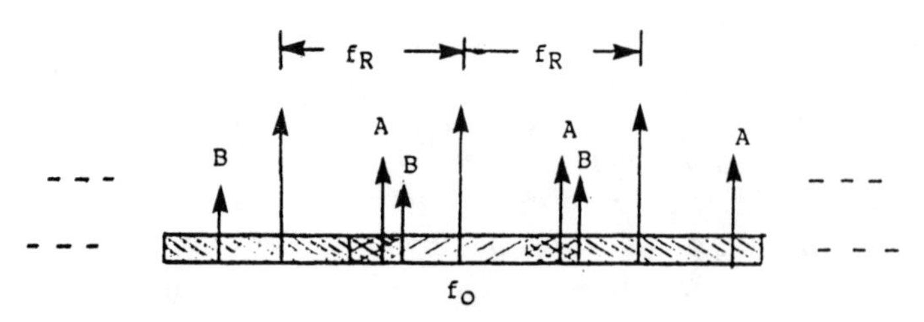

Figure 5.6 CW and pulsed radar doppler spectrums.

frequency f_0, the doppler line spectrum of an approaching target at frequency f_A, and the doppler line of a receding target at f_B. The crosshatched band of width B_d represents the range of possible doppler frequencies:

$$B_d = 2f_{d\max}$$

This spectrum, analyzed by a doppler filter bank or a spectrum analyzer, will reveal the three spectral components (f_0, f_A, and f_B) in their true locations with no ambiguity. Figure 5.6(b) shows the spectral content of the RF return of a coherent pulsed radar observing these same targets. The pulsing amounts to a multiplication of the target return by the repetitive pulse train with spectrum consisting of lines that are at multiples of the PRF f_R. The RF spectrum now contains the lines at f_0, f_A, and f_B, within the original band of width B_d centered on f_0, along with replicas of this spectrum, centered on the pulse spectrum lines at $f_0 \pm nf_R$, where n is an integer.

A filter of width B_d centered on f_0 or on any of the pulse spectrum lines captures all the doppler spectrum information that was available in the CW line spectrum of Figure 5.6(a). The pulsed radar permits the option of setting a range gate at some selected range in order to observe returns from that specific range. If targets A and B lie within the range gate, the spectrum of the gated output will be just as described. Once the range gating has occurred, spectrum analysis of any of these bands of width B_d will reveal the doppler content of returns from the selected range. As is clear in Figure 5.6(b), if the PRF is gradually reduced, the spacing between spectral bands will shrink, and eventually, when f_R reaches a value B_d, the spectral bands will begin to overlap. This results in spectral "aliasing" as in Figure 5.6(c), in which the true doppler of an observed spectral component can no longer be ascertained. The width B_d was equal to $2f_{d\max}$, so, aliasing or spectral ambiguity can be avoided if the PRF is at least:

$$f_{R\min} = 2f_{d\max}$$

5.4 THE PRF RANGES OF AIRBORNE COHERENT RADARS

Modern airborne military radars are multimode devices with computer-controlled operating modes to match the aircraft's mission. A terminology has evolved for categorizing the modes of these radars according to PRF (the same categorization can be applied to surface-based radars). The three PRF ranges are defined as follows:

Low-PRF mode: unambiguous range, highly ambiguous doppler;
Medium-PRF mode: both range and doppler moderately ambiguous;
High-PRF mode: unambiguous doppler, highly ambiguous range.

Some signal-processing functions in low-PRF radars use the doppler phenomenon (e.g., MTI), but the low sampling rate makes doppler so ambiguous that it is not one of the measured signal parameters. In the medium- and high-PRF modes, target doppler is measured and may be considered one of the radar signal (target) coordinates (Appendix B). Thus, only the medium- and high-PRF modes are called *pulsed doppler modes.* Doppler measurement can be implemented in analog form, but most modern pulsed doppler radars employ digital signal processing.[3] Processing for doppler involves many consecutive samples of the target return, collected over a time span called a *coherent processing interval* (CPI).

[3]Sometimes the term *signal processing* is reserved for analog processes, and the term *data processing* is reserved for digital processors.

Low-PRF Mode

The low-PRF mode is useful for ground mapping for navigation or bomb delivery. For "real-beam" ground mapping,[4] doppler processing may not be used at all, as is the case when analog envelope-detected video is displayed directly. For detecting slow-moving surface targets, the only doppler processing may consist of clutter suppression by an MTI notch filter. We may implement range-gated doppler filtering to provide clutter rejection and coherent integration of moving-target returns. This arrangement requires many contiguous range gates to cover the large unambiguous surveillance range of which the low-PRF radar is capable.

The SAR mode of ground mapping yields azimuth resolution that is much finer than is possible in the real-beam mode. SAR processing requires unambiguous doppler measurement within the doppler band corresponding to the spread in doppler[5] observed in the main-lobe terrain returns. This spread is given by

$$\Delta f_d \approx \frac{2V}{\lambda} \beta \sin \theta$$

where v is the aircraft velocity, β is the antenna azimuth beamwidth, and θ is the azimuth "squint angle" of the beam relative to the velocity vector. For SAR mapping at $v = 100$ m/s, $\lambda = 0.03$ m, $\beta = 0.05$ rad, and $\theta = 60°$ the doppler spread across the beam is only

$$\Delta f_d = 290 \text{ Hz}$$

Therefore, the low-PRF mode clearly is compatible with SAR map processing.

High-PRF Mode

To understand the PRF requirement for the high-PRF (unambiguous doppler) mode for the airborne look-down radar, let us return to the situation described by Figure 5.2. Figure 5.7(a) shows the sidelobe clutter spectrum extending from

$$f_{c\max} = \frac{2v_P}{\lambda}$$

corresponding to clutter returns from dead ahead, to

[4]As distinguished from *synthetic aperture radar* (SAR) mapping.

[5]Because we can remove the absolute doppler at beam center, we need be concerned only about the spread about beam center in setting the sampling rate.

$$f_{c\min} = -\frac{2v_P}{\lambda}$$

corresponding to clutter returns received in the antenna backlobes. The main-lobe clutter peak location depends on the main-lobe scan angle. The peak will fall between $f_{c\max}$ and

$$f_{c\theta} = \frac{2v_P}{\lambda} \cos\theta$$

where θ is the scan angle. If the range bin (resolution cell) under consideration[6] encompasses the area directly below the aircraft there will be an "altitude line" at zero doppler. This peak rises above other sidelobe clutter because the range is short and, primarily, because the backscatter coefficient σ° becomes very large at normal incidence. Target doppler, as noted earlier (refer to Figure 5.1 and (5.4)), is given by

$$f_T = \frac{-2}{\lambda}(v_{rT} - v_{rP})$$

For a head-on encounter, the magnitude of v_{rT} is equal to the total target velocity v_T, but it carries a negative sign:

$$v_{rT} = -v_T$$

so the maximum target doppler ever encountered is

$$f_{T\max} = \frac{2}{\lambda}(v_{T\max} + v_{P\max})$$

The target spectrum at f_T in Figure 5.7(a) represents a target closing (approaching) fast enough to place its doppler in the clear region beyond the sidelobe clutter spectrum. Figure 5.7(b) is the corresponding pulse spectrum. The portion of the spectrum extending from $-f_R/2$ to $+f_R/2$ is labeled *central band.* This band will be processed for detection and tracking. In this illustration, f_R is sufficiently high that there is no aliasing of receding sidelobe clutter from the next spectral band into the central band. To provide the maximum clear (clutter-free) spec-

[6]In the high-PRF mode there may be only a single range bin. Even with multiple range bins, the range ambiguity problem may cause the "altitude return" to fold over into any range bin.

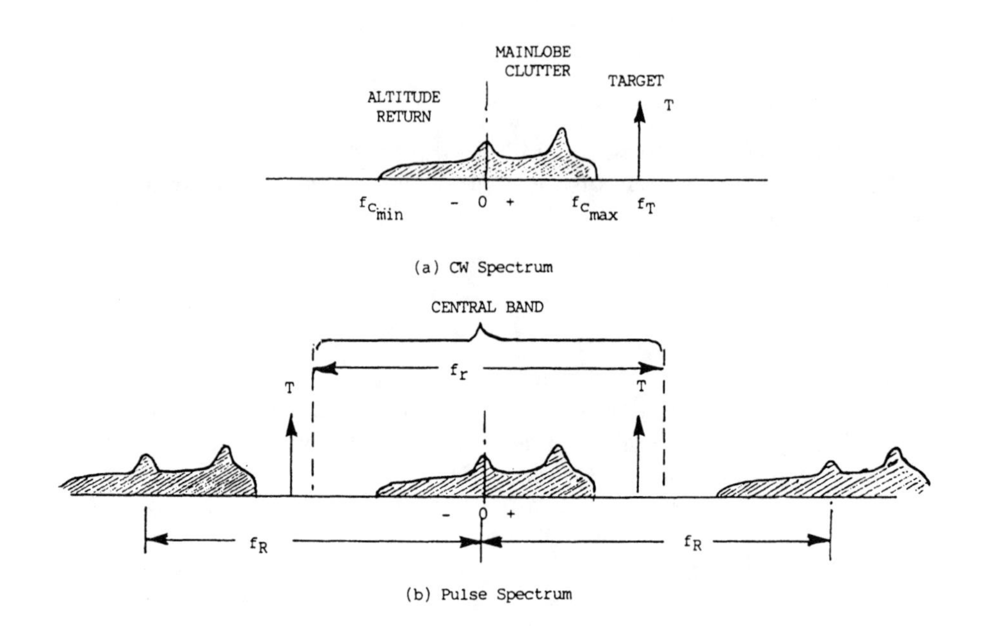

Figure 5.7 PRF requirement for avoiding doppler ambiguities.

trum space for detection of closing targets, f_R must be high enough to ensure that all receding sidelobe clutter from the next spectral band will lie beyond $f_{T\max}$. This requires

$$f_R \geq f_{T\max} + |f_{c\min}|$$

or

$$f_R \geq \frac{2}{\lambda}\left[v_{T\max} + v_{P\max} + v_{P\max}\right] = \frac{2}{\lambda}\left[v_{T\max} + 2v_{P\max}\right] \qquad (5.5)$$

For the subsonic flight example mentioned earlier ($v_{P\max} = v_{T\max} = 330$ m/s), with $\lambda = 0.03$ m, the minimum PRF required by (5.5) is $f_R \geq 66$ kHz. For supersonic flight, and for radars operating at shorter wavelengths, the PRF requirement can reach several hundred kHz. For the tail-chase situation, the target doppler can be low, approaching zero when the two aircraft velocities become equal, and becoming negative if the pursuer is losing the chase. Thus, target doppler in the tail chase falls within the sidelobe clutter spectrum.

It is clear from (5.5) that the high-PRF mode can yield a great many received pulses per beam dwell period. If, for instance, the antenna's azimuth scan rate is $\Omega_s = 100°/\text{s}$ and the azimuth beamwidth is $\beta = 3°$, the beam dwell duration is

$$T_\beta = \frac{\beta}{\Omega_s} = 0.03 \text{ s}$$

The number of target return pulses received during the dwell is

$$n_\beta = f_R T_\beta$$

For the 66-kHz PRF, $n_\beta = 1980$. If the entire set were processed in a single CPI, the attainable doppler resolution would be, approximately,

$$\Delta f_d = \frac{1}{T_\beta} = 33 \text{ Hz}$$

This corresponds to a velocity resolution of

$$\Delta v = \frac{\lambda}{2} \Delta f_d = \frac{0.03}{2} (33) \approx 0.5 \text{ m/s}$$

This may be considerably finer than required, so the total dwell may be divided into several CPIs. This simplifies the processing. We can then sum the outputs (integrated noncoherently) of the several CPIs. It is likely that instead of holding f_R constant for the entire dwell, different values of PRF would be used for each CPI. This permits the range-eclipsing problem to be solved.

In the medium-PRF mode, the use of multiple PRFs also provides a way of solving the range ambiguity problem, but in the high-PRF mode only a single range bin, or at most a few bins, exists, so another way of determining range is employed when target range is required. Eclipsing occurs when the target is in a range band from which echoes arrive in time coincidence with transmission of a pulse (reception is inhibited during that time). These eclipsed (blind range) bands are spaced by the range ambiguity interval, ΔR_{ambig}:

$$\Delta R_{\text{ambig}} = \frac{c}{2f_R}$$

For $f_R = 66$ kHz, $\Delta R_{\text{ambig}} = 2.27$ km. For a radar with a maximum range of 100 mi (160 km) there are

$$n_R = \frac{R_{\max}}{\Delta R_{\text{ambig}}} = \frac{160}{2.27} = 70$$

range ambiguity intervals. The fraction of each interval that is eclipsed is equal to the transmission duty factor, DF_X, of value:

$$\mathrm{DF_X} = \frac{T}{T_R} = Tf_R$$

where T is the duration of the transmitted pulse. For the high-PRF mode, $\mathrm{DF_X}$ can be from 0.25 to 0.5. Thus, in 25% to 50% of the situations, a target lies within a blind range band. For the previous example, if $\mathrm{DF_X} = 0.5$ then the width of a blind range band is

$$\Delta R_{\mathrm{blind}} = \mathrm{DF_X}\, \Delta R_{\mathrm{ambig}} = 0.5 \times 2.27 = 1.14 \text{ km}$$

We do not want lapses in detection or tracking of the duration equal to the several seconds that are required for a target to fly through a blind range band. Therefore, we alter the PRF from one CPI to the next in a pattern that shifts the eclipsed bands to ensure that eclipsing never persists for very long. If the parameters are such that only a single CPI can be accommodated per dwell, the PRF can be changed from scan to scan.

In the high-PRF mode, the pulse duration $T = \mathrm{DF_X}/f_R$ is not large. For instance, if $f_R = 10^5$ Hz and $\mathrm{DF_X} = 0.25$, the pulse duration is only $T = 2.5\ \mu$s. Therefore, it is common practice to use simple pulses, for which $BT = 1.0$, B being the pulse bandwidth. The range resolution of the simple pulse is

$$\Delta R_{\mathrm{resol}} = \frac{C}{2B} = \frac{CT}{2} = \mathrm{DF_X}\, \Delta R_{\mathrm{ambig}}$$

The width of the uneclipsed portion of the range ambiguity interval is

$$\Delta R_{\mathrm{uneclipsed}} = \Delta R_{\mathrm{ambig}}(1 - \mathrm{DF_X})$$

Therefore, the number N_B of possible range bins (resolution elements) within the uneclipsed band is

$$n_B = \frac{\Delta R_{\mathrm{uneclipsed}}}{\Delta R_{\mathrm{resol}}} = \frac{1}{\mathrm{DF_X}} - 1$$

For the duty factor values mentioned, this amounts to only two or three possible range bins. To use a single range bin (no range gating) is not uncommon, in which case only a single doppler filter bank is required.

The high-PRF mode, with its high average power (high transmit duty factor) and its clutter-free doppler region, provides the best means for detecting approaching targets in what is called the *velocity-search* mode. Detections in this mode are normally displayed on a velocity *versus* azimuth display, the range being unknown. When this mode reveals a potentially threatening target, a transition to a *range-*

while-search mode is initiated. In this mode, the high PRF is retained, but alternating with the CPIs of constant carrier frequency are CPIs in which the carrier frequency is linearly swept. This is the counterpart of FM ranging in the CW radar. The range resolution attainable is given by

$$\Delta R = \frac{c}{2\,\Delta f}$$

where Δf is the frequency excursion of the linear FM sweep during the CPI.

Medium-PRF Mode

Airborne radars with a look-down capability generally have a medium-PRF mode in which a set of PRFs is available, perhaps from 10 to 30 kHz. The medium PRF yields a higher detection probability in the tail-chase situation [2, Chapter 6], and provides accurate range information directly from pulse delay measurement, whereas in the high-PRF mode only FM ranging of lower accuracy is available. As we already noted, when the PRF is reduced from the high (unambiguous doppler) value, the signal and clutter spectral bands centered on PRF harmonics overlap: the lower the PRF, the greater the multiplicity of spectral foldovers. In the medium-PRF mode, there is no clutter-free space in the central band of the spectrum that is processed; sidelobe clutter fills the spectrum. Regions of the processed band (of width equal to the PRF) containing main-lobe clutter or altitude return are blanked; that is, the outputs of the doppler filters containing those strong clutter components are ignored. Those blanked frequencies are "blind" doppler frequencies. Some members of the available PRF set are chosen specifically so that the blind doppler bands can be shifted from one CPI to the next to ensure that no target will remain long in a blind doppler situation. There are also blind ranges corresponding to the eclipsed ranges already discussed. The fraction of the total range expanse that is eclipsed is much smaller than in the high-PRF mode, because DF_X is smaller. The fraction would be extremely small if narrow simple pulses were transmitted, but, generally, in this mode pulse-compression waveforms are employed to achieve higher average power and, consequently, better detection probability than would be possible with a simple short pulse.

The lower PRF (longer PRI) makes time available for the longer pulse-compression waveform. Even with the less severe range-eclipsing problem, the medium-PRF mode requires PRF changes from one CPI to the next to shift the blind ranges to ensure that no target remains long in a blind range zone. Depending on the scan rate (and associated target dwell duration), to cycle through the different PRFs (one for each CPI) during a single dwell may be possible. If not, some members of the PRF set can be used on one scan, and the others on the next. Five to 10 PRFs may be required to avoid blind ranges and blind dopplers and to resolve the range and doppler ambiguities that exist in this mode. Range ambiguity resolution is less difficult than in the high-PRF mode because there are fewer range foldovers. However, this advantage is bought at a cost of the moderate level of doppler foldover that comes with the lower PRF.

5.5 PULSED DOPPLER SIGNAL PROCESSING

The flowchart of Figure 5.8 shows the signal-processing functions that would be involved in a pulsed doppler radar that has both medium- and high-PRF modes. In the high-PRF mode, it is likely that a simple pulse would be transmitted, so the pulse-compression filter would not be used. Moreover, it is likely that there would be a single range gate, so only a single doppler filter bank would be required. Because Figure 5.8 presumes monopulse angle tracking, the channels Σ, Δ_{az}, and Δ_{el} are carried throughout. Actually, a pair of channels can suffice in the digital-processing section. One member of the pair would always carry the Σ information, and the other member would be time-shared on alternate CPIs between Δ_{az} and Δ_{el}. The Σ channel output must always be available for demodulating the difference outputs to derive the angle error signal Δ/Σ (see Chapter 4). If conical scan angle tracking were employed, a single receiver and processor channel would replace the three channels of the monopulse system.

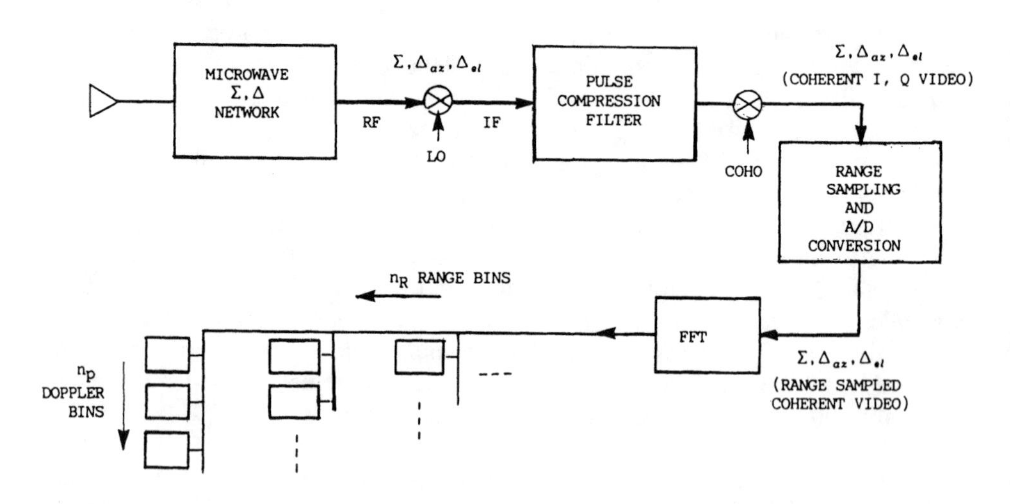

Figure 5.8 Pulsed doppler processing flowchart.

The FFT block is a fast-Fourier-transform processor. The input is the n_p pulse samples collected from a given range gate during the CPI (the coherent integration time). The output is the discrete Fourier transform of that input, i.e., the realization of the doppler filter bank for that range gate (range bin). The process is replicated for all the range gates. The coherent integration time (CPI duration) is

$$T_i = n_p T_R = \frac{n_p}{f_R}$$

The resulting doppler resolution is

$$\Delta f = \frac{1}{T_i} = \frac{f_R}{n_p}$$

The process yields n_p output samples of the signal's doppler spectrum. These outputs can be viewed as n_p continguous doppler filters, each of bandwidth Δf, with a spacing Δf between filter center frequencies. The unambiguous doppler frequency band covered by the bank of n_p filters is

$$B_d = n_p \Delta_f = f_R$$

The FFT size (value of n_p) is set equal to a power of 2. In the high-PRF mode, typical values might be $n_p = 2^9 = 512$ to $n_p = 2^{11} = 2048$. In the medium-PRF mode, the values might be $n_p = 2^4 = 16$ to $n_p = 2^6 = 64$.

Section 1.7 discusses signal-processing gains and their effect on J/S requirements. The processing of Figure 5.8 includes both the single-pulse processing gain of the pulse compression and the n_p-pulse coherent integration gain inherent in the FFT processor. In the medium-PRF mode, the pulse-compression ratio,[7] BT, would probably have values in the range:

$$10 \leq BT \leq 100$$

producing processing gains of

$$10 \text{ db} \leq g_{p1} \leq 20 \text{ dB}$$

In the n_p-pulse coherent integration process, the gain is n_p itself. For the n_p values suggested earlier, for the high-PRF mode the processing gain would be

$$27 \text{ dB} \leq g_{pn} \leq 33 \text{ dB}$$

for the medium-PRF mode:

[7] B is the pulse bandwidth, and T is the duration of the uncompressed pulse (see Section 1.7).

$$12 \text{ dB} \leq g_{pn} \leq 18 \text{ dB}$$

Additional processing gain can accrue when the available time permits several CPIs at the same PRF. The outputs of m coherent integrations can be added (noncoherently integrated) for a noncoherent integration gain, approximated in Section 1.7 by

$$g_{pm} = m^{0.8}$$

As an illustration, assume the following parameters for the medium PRF mode.

Example:

β = antenna azimuth beamwidth = 3°;
Ω = azimuth scan rate = 100°/s;
f_R = PRF[8] = 15 kHz;
BT = pulse-compression ratio = 10;
n_p = size of FFT = 2^5 = 32;
T_β = dwell on target = β/Ω = 30 ms;
T_i = average CPI duration = n_p/f_R = 2.1 ms.

Let us assume that four different PRFs will suffice. There is then sufficient time during the dwell for $m = 3$ CPIs at each PRF:

$$4 \text{ PRFs} \times 3 \text{ CPIs/PRF} \times T_i = 12T_i = 25 \text{ ms}$$

so there are 5 ms left over for transmission of pulses for determining AGC settings and for other housekeeping functions. The processing gains are

$$g_{p1} = 10 \log(BT) = 10 \text{ dB}$$
$$g_{pn} = 10 \log(n_p) = 15 \text{ dB}$$
$$g_{pm} = 10 \log(m^{0.8}) = \underline{3.8 \text{ dB}}$$
$$\text{Total } 28.8 \text{ dB}$$

Pulsed Doppler Detection

The range-gated, doppler-filter processing of Figure 5.7 yields, upon completion of processing of all range gates, an output signal array depicted by Figure 5.9. There

[8]Several different PRFs will be selected to resolve ambiguities and to avoid blind ranges (eclipsing) and blind dopplers (filters that are filled with main-lobe clutter). This value of PRF is an average of the several values.

are n_r range gates and n_p doppler filters per range gate, for a total of $n_r n_p$ range doppler resolution cells. Some regions of this range doppler array may be blanked (that is, ignored)—for example, the regions containing main-lobe clutter and perhaps attitude line clutter. The remainder is searched for targets. A target return may produce responses in more than one resolution cell. This is the case when the target straddles the boundary between cells or when the resolution in either or both dimensions is finer than the width of the target response. The doppler filter formation process results in filters with finite doppler sidelobe response. Pulse compression, if employed, results in finite-range (time) sidelobes.

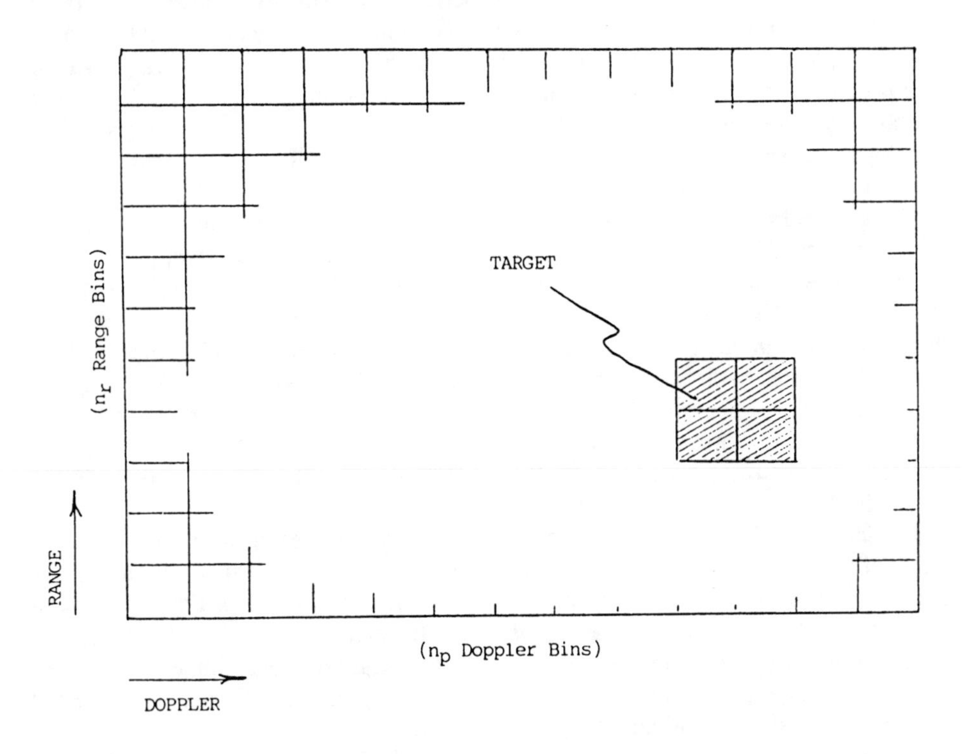

Figure 5.9 Pulsed doppler processor output array.

In Figure 5.9, a target detection is indicated in the shaded four-cell block. This means that a single target has produced responses that are above the detection threshold in all four cells. An interpolation process, based on the relative amplitudes in the signal levels in the four cells, can lead to a detection declaration in one of the four cells or to a more precise indication of range and doppler, measured to fractions of cell widths. With monopulse angle processing, Figure 5.8 would be replicated[9] for Σ, Δ_{az}, and Δ_{el} channels. From these, one could develop the elevation and azimuth error signals, thereby yielding estimates of the angle coordinates of the target to within a fraction of a beamwidth. The process would be repeated again and again as the antenna proceeds through an azimuth scan. Possibly, other targets would be detected at other azimuths.

Pulsed Doppler Tracking

If a target were to be engaged in an attack, perhaps by an air-to-air missile, the radar might transition to a *single-target tracking* (STT) mode. The scan would be shrunk to a small, acquisition sector scan about the target azimuth, and the range gating would be reduced to a few cells centered on the target range. Moreover, knowing the target doppler, we might alter the FFT parameters, but fine doppler resolution and high integration gain would still be required. Finally, upon completion of acquisition, the antenna scan would be halted at the target azimuth and automatic tracking would begin. Updating of angle, range, and doppler estimates would then occur once per CPI or once every few CPIs. Once we know the target parameters, we can choose a single PRF or a pair of PRFs to avoid blind range and blind doppler situations. We would have to change the PRF from time to time, however, as the target coordinates change. The filtering, by the coordinate track loops, of these frequent samples leads to accurate tracking in the STT mode.

To continue search while tracking, we need to reduce the tracking sample frequency to one sample per scan. This becomes a TWS mode. In a radar with a mechanically scanned antenna, the TWS mode provides the only means for tracking multiple targets. For range tracking, we can select a pair of cells to serve as early or late gates (see Chapter 3). This could be pairs from either column of the shaded block of four cells in Figure 5.9. We probably would choose the pair containing the strongest signal. For doppler tracking, a pair of cells from one of the two rows of the shaded block would be selected as the two adjacent (high and low) doppler filters for deriving the doppler error signal. As the target moves relative to the radar, the shaded area representing the target range and doppler coordinates in Figure 5.9 will shift its position. For each CPI the selection of the cell pair for developing the error signal in each coordinate has to be reviewed. For monopulse angle tracking, we must choose a cell (probably the cell with the strongest Σ signal) for developing the angle error signal, Δ/Σ. Now, n_p pulses are involved in arriving at a single angle estimate, whereas in the conventional (nondoppler) radar a monopulse angle estimate is derived on each received pulse.

For conical scan, or lobe switching in two angle coordinates, a minimum of one complete scan or lobing cycle is required to develop target angle estimates (see

[9]Only the Σ channel may be employed during the detection phase.

Chapter 4). For conical scan at a 20-Hz scan frequency, 1/20 s is required. If, as in the previous example, T_i = 2.1 ms, then 24 CPIs would transpire during one complete conical scan cycle. To derive the conical scan amplitude modulation, perhaps one would average signals from all the cells containing target contributions above threshold (the shaded cells in Figure 5.8).

5.6 PULSED DOPPLER VELOCITY DECEPTION

The requirements for velocity deception against a pulsed doppler radar are basically the same as those described in Section 5.2 for velocity deception against a CW radar. Both require transmission of a carrier with a false-doppler shift that is gradually pulled away from the doppler of the target's skin return. The difference is that against the CW radar the jammer should transmit continuously, whereas against the pulsed doppler radar the jammer should respond with pulses.

In principle, a CW jammer with the proper VGPO modulation of the carrier frequency will inject the same form of signal into the central band of a pulsed doppler spectrum (Figure 5.6), as would a pulsed jammer. Likewise, the fundamental spectral component of a pulsed jammer spectrum will look no different to a CW radar receiver than would a CW jammer signal of the same carrier frequency. There are two reasons for using CW jamming against the CW radar and for using pulsed repeater jamming against the pulsed doppler radar. The first is efficiency. Only that portion of CW jamming received during the pulsed radar's "listening" period is effective; the rest is wasted. As we noted in the discussion of the high-PRF case, listening may occupy as little as 25% of the time. On the other hand, if pulsed jamming is used against a CW radar, only the fundamental component of the pulse spectrum energy is useful. The amplitude of the fundamental component is equal to the pulsed carrier amplitude multiplied by the pulsed transmit duty factor. The other reason for matching the form (CW or pulsed) of the jamming to the victim radar's mode is that a smart radar might more readily recognize the wrong form of jamming as actually being jamming and take measures to avoid VGPO. For instance, the radar might override its AGC to prevent a strong jammer from driving the receiver gain so low that the skin return would be undetectable. It might then take steps to ensure that its doppler tracker stays with the skin-return signal, ignoring the jammer as it executes its VGPO maneuver.

5.7 VELOCITY-DECEPTION SIMULATION RESULTS

Section 5.5 describes doppler tracking on a TWS basis, thus enabling the tracking of multiple targets. We could make the range-gated doppler filter bank of Figure 5.8 a part of an STT mode, but we would use only a pair of doppler bins to develop the doppler error signal. This error signal would control the frequency of a voltage-tunable local oscillator to keep the target doppler of interest centered on the pair of doppler bins. Outputs of all the other bins would be ignored.[10]

For the STT mode, the filter bank is not needed; we can replace the single pair of cells used for doppler error detection by a frequency discriminator. The doppler trackers with which we are concerned in this section are often called *velocity gates* or *speed gates* because they automatically keep the doppler spectrum of a single target centered in a narrow-band filter that is, in fact, a doppler gate.

We discussed earlier a CW doppler tracker that had only a single spectrum line to deal with. In Figure 5.10 we consider a pulsed doppler radar receiver. The IF section, centered at f_{i1}, has a bandwidth matched to the pulsed signal bandwidth. Range gating is performed in this IF path to select the target. The range-gated return is then translated to a lower center frequency, f_{i2}, and passed through a narrow-band filter (the velocity gate). This filter is just wide enough to pass the central line of the pulsed doppler spectrum of the target return. The line has a finite width. A small amount of broadening is due to target scintillation and time-varying multipath effects. In addition, the return may contain angle information (as in CONSCAN and lobe switching). It may carry range information if the range tracker employs audio-frequency dither of the range gate (this is analogous to lobe switching in angle; it enables a range tracker to track on doppler-filtered modulation sidebands from a single range gate rather than requiring a split-gate pair of range-gated doppler channels). Thus, the velocity-gate filter must be wide enough to pass any modulation sidebands needed by the angle- and range-tracking loops. The matching of filter width to the target spectrum yields the maximum S/N and *signal-to-clutter* ratio (S/C) attainable in the track loops. Target doppler can vary greatly, so the target doppler must first be pulled into the velocity-gate passband (this is the acquisition process), and it must then be held there by the doppler tracking loop.

In Figure 5.10, the doppler tracker depends on a frequency discriminator that senses any shift of the target doppler off the center frequency f_{i2}. The error voltage from the discriminator pulls the frequency of the VCO in the proper direction to recenter the target spectrum on f_{i2}. Thus, the loop is an AFC loop. The loop filter is designed to provide the necessary transient and steady-state control characteristics (see Appendix A). For instance, if the filter contains a single integrator (type 1 loop), the loop can track a steady doppler with zero-frequency error, and if the doppler varies linearly with time (target acceleration is constant) there will be a finite tracking error.

We can replace the AFC loop by an automatic phase control loop commonly called a PLL. The frequency reference (the discriminator) must be replaced by a phase reference (a stable oscillator). In Figure 5.11 the doppler tracker is imple-

[10]The other bins could be monitored for other targets in the same range bin, moving at slightly different speeds, or for target modulation spectrum components.

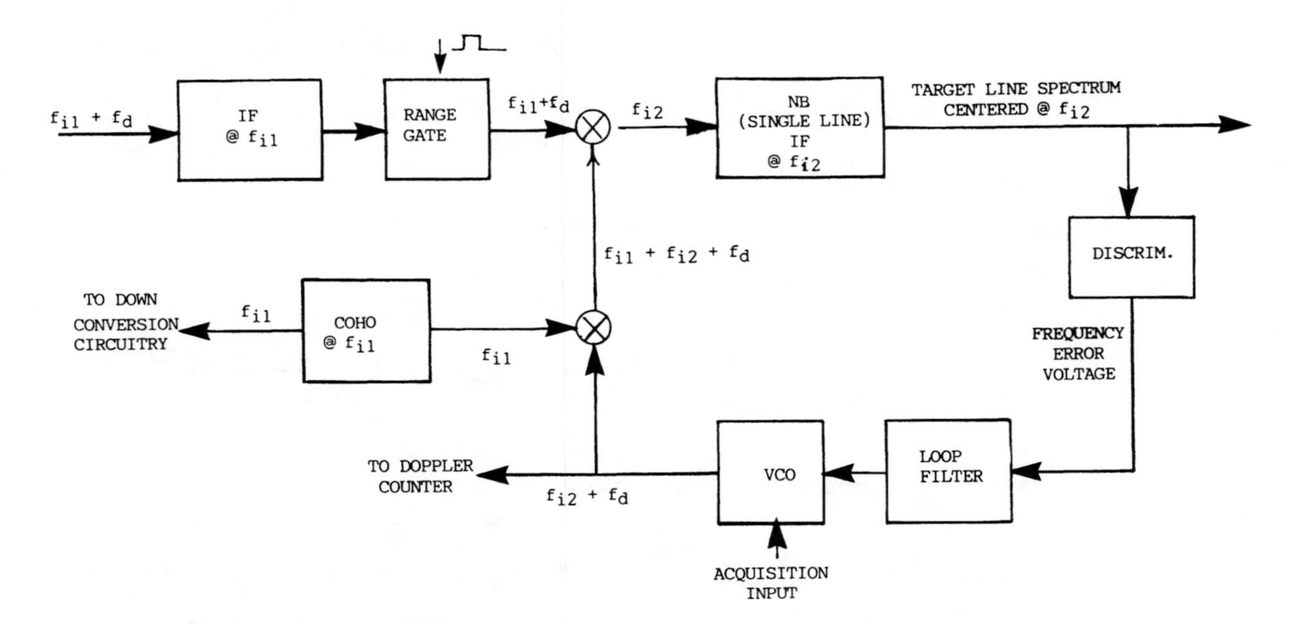

Figure 5.10 A doppler tracker implementation.

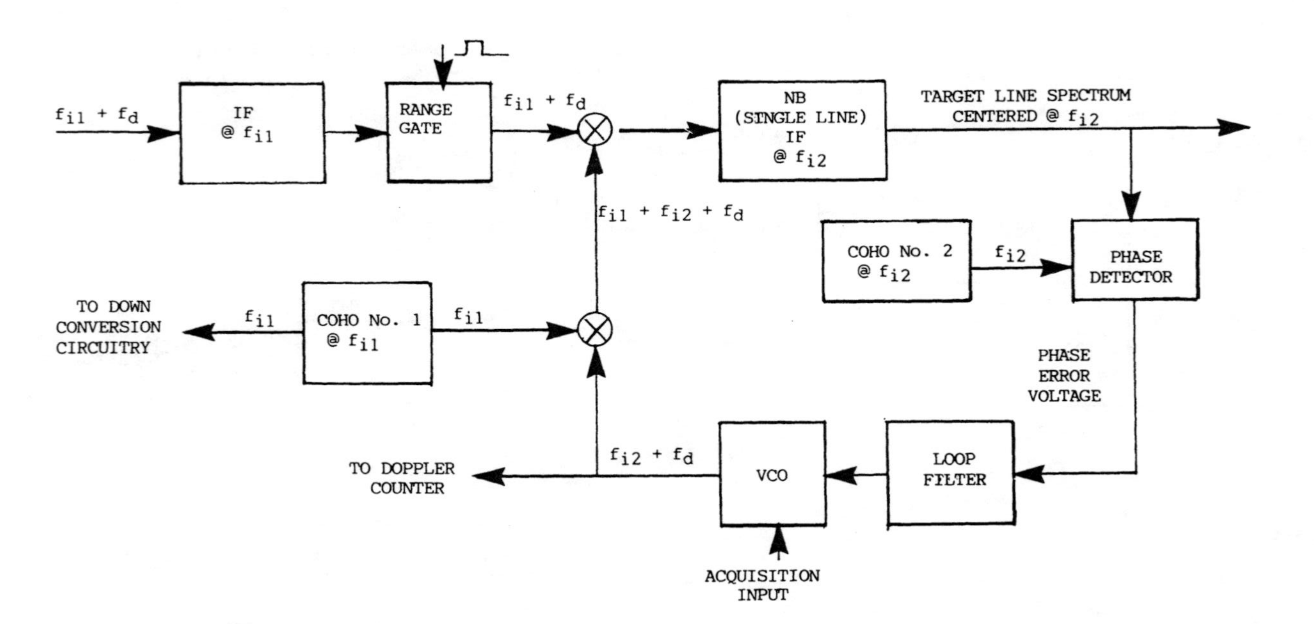

Figure 5.11 Doppler tracker using phase-locked loop.

mented as a PLL by using the COHO No. 2 oscillator as the phase reference.[11] In both schemes, the VCO tracks the target doppler by maintaining its frequency at $f_{i2} + f_d$, where f_d is the target doppler. Thus, a counter reading the VCO frequency provides a readout of target doppler (by automatically subtracting f_{i2} from its count).

Because the VCO frequency is proportional to the control voltage, the VCO output phase becomes proportional to the integral of the control voltage. To the extent that the phase detector yields an error voltage that is linearly related to phase error, the loop is a type 1 loop if the filter contains no integrator. If the filter contains an integrator, the loop becomes type 2. The type 1 loop tracks with a finite phase error when input phase is varying at a constant rate (a constant f_d). The type 2 loop tracks with a finite phase error if the input phase *rate* is increasing (or decreasing) linearly with time (as in a target moving with constant acceleration). Because the doppler tracker has a very narrow passband, the signal must initially be acquired either through a doppler search (frequency scan of the VCO) or through acquisition information passed to the loop from an outside source (acquisition input in the block diagrams).

If a jammer is to execute a VGPO maneuver, it must begin by offering a jamming frequency equal to the doppler-shifted target return frequency being tracked. With a repeater jammer aboard the target vehicle, the jamming frequency automatically has the desired initial value. After allowing the radar AGC time to adapt to the level of the jamming (which will normally be higher than the level of the target return), the repeater needs to impress a time-varying frequency offset on its output to lure the doppler tracker away from the target return. If the jammer is a transponder with frequency not identical to the target return (but presumably very close), the jammer must delay its pull-off for a time long enough to allow the doppler tracker to drop the target and acquire the stronger jamming signal.

To illustrate the VGPO problem, we simulated a simple doppler tracker, using a PLL containing a loop filter with transfer function:

$$G_f(s) = \frac{K(sT + 1)}{s}$$

Accounting for the integration inherent in the VCO, we write the overall open-loop transfer function as

$$G(s) = \frac{G_f(s)}{s} = \frac{K(sT + 1)}{s^2}$$

[11]COHO is an abbreviation for *coherent oscillator*. Note that COHO No. 1 was involved in the heterodyne process that produced f_{i1}, the first IF center frequency.

This is the small-error linear approximation.[12] The simulation model contains a phase detector with characteristic:

$$V_\epsilon = \sin(\phi_\epsilon)$$

where ϕ_ϵ is the phase error, so the linear approximation is good as long as the error is only a fraction of a radian. The constant, K, of the transfer function becomes the steady-state acceleration error constant (see Appendix A). Thus, if we assume that linearity holds for a constant phase acceleration, $\ddot{\phi}$, of the target (real or false), the steady-state phase tracking error is

$$(\epsilon_\phi)_{ss} = \frac{\ddot{\phi}}{K} = \frac{1}{K}\left(\frac{d\omega_d}{dt}\right) = \frac{1}{K}\left(\frac{4\pi}{\lambda}\frac{dv}{dt}\right)$$

where ω_d is the doppler radian frequency (rad/s) and v is radial velocity. Thus, if the tracker is deceived into accepting a false doppler, ω_d, which is then pulled off at rate $\dot{\omega}_d$, the tracker should follow with an error:

$$(\epsilon_\phi)_{ss} = \frac{\dot{\omega}_d}{K}$$

The pull-off must be slow enough for the tracker to follow. For example, if $(\epsilon_\phi)_{ss}$ approaches $\pi/2$, the jammer is on the verge of losing the tracker. (This represents the peak of the $\sin(\phi)$ error detector curve.) In the simulations, the gain constant was set at $K = 10^4\ s^{-2}$. This results in a closed-loop tracker bandwidth of

$$B \approx \sqrt{K} = 100 \text{ rad/s}$$

No attempt was made to relate pull-off maneuvers to specific target accelerations or radar wavelengths. We might observe, however, that a value $\dot{\omega}_d = 1000$ rad/s^2 corresponds to an acceleration of

$$\frac{dv}{dt} = \frac{\lambda}{4\pi}\dot{\omega}_d = 80\lambda \text{ m/s}$$

so at $\lambda = 0.1$ m the acceleration would represent 8 m/s^2 or about $0.8g$ (where g is the acceleration of gravity).

[12]The value of T was selected to yield only moderate overshoot in the step response.

In the simulation results shown in Figures 5.12 through 5.14, the pull-off maneuver is a constant rate of change of doppler (constant $\dot{\omega}_d$).[13] The three cases shown differ in the values of J/S and the behavior of the jammer before initiation of the pull-off. In the plots, time is measured in seconds. The variables ϕ_j and ϕ_v (labeled PHEJ and PHEV) are, respectively, the jammer phase and the VCO phase, measured in radians, relative to the target phase (after the phase of the COHO is

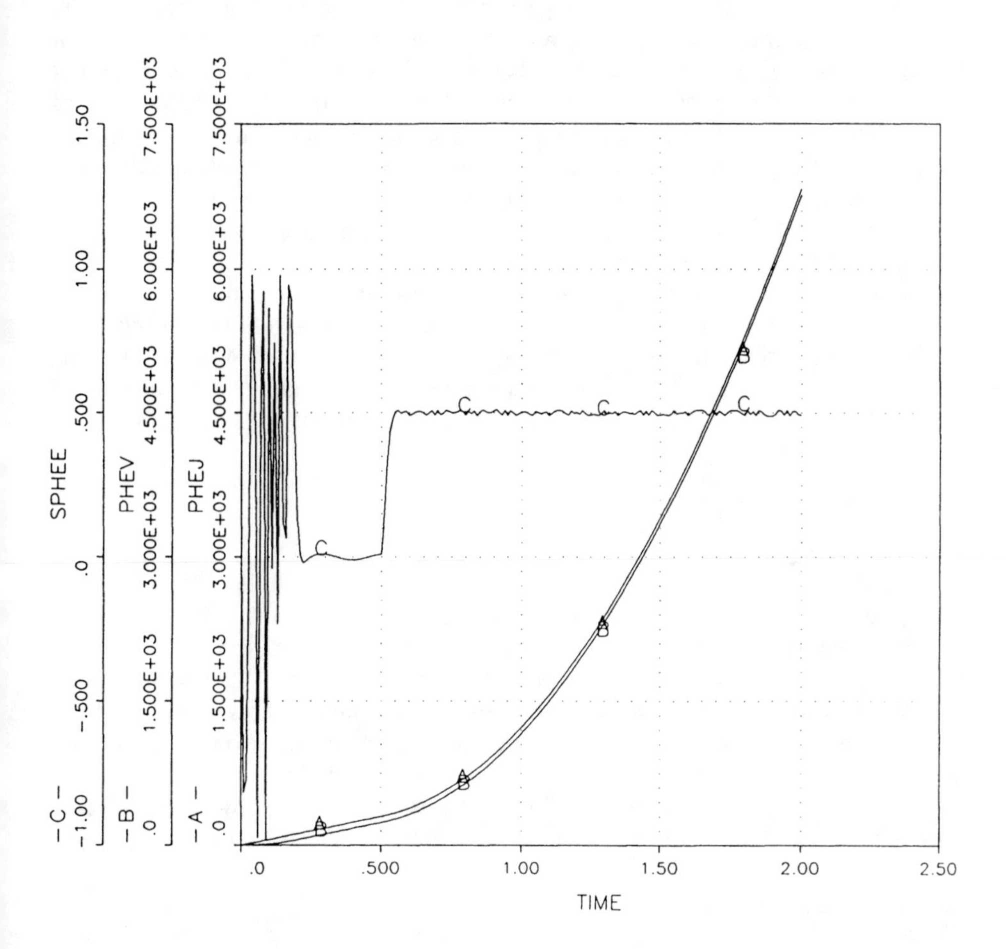

Figure 5.12 VGPO simulation with small frequency set-on error.

[13]Note that this is an unrealistic maneuver for a real target; it represents an acceleration step. An intelligent tracker would very likely recognize this as a false target.

removed). These variables would not normally be displayed in an actual tracker. Instead of displaying phase error ϕ_ϵ, the output of the phase detector, $\sin(\phi_\epsilon)$, is displayed; it is labeled SPHEE. This is a convenient variable to observe. When the loop loses phase track and slips cycles, $\sin(\phi_\epsilon)$ cycles between ± 1.0.

In Figure 5.12 there was an initial jammer set-on error of 600 rad/s. The jammer held this set-on frequency for 0.5 s and then began its pull-off at a rate of $\dot{\omega}_d = 5000$ rad/s^2. The J/S was very high (40 dB).[14] As the VCO goes through the transient period in which it shifts from the target to the jammer, it slips cycles until it attains a phase lock after about 0.2 s. Because of the finite granularity of the plot point spacing, not all these cycles are observable as ± 1.0 cycling of the value of $\sin(\phi_\epsilon)$. The number of cycles slipped is unimportant, but lock-on time is a matter of concern to the jammer programmer. Because ϕ_j and ϕ_v are true phase values (all multiples of 2π are included in the phase count), the vertical separation between the ϕ_j and ϕ_v plots includes not just the modulo 2π value of ϕ_ϵ but the total value, including all the cycles slipped during lock-on. This separation in Figure 5.12 appears to be about 70 rad, representing a slipping of perhaps 10 or more cycles (note that lock-on time does not reveal cycle count, for the VCO frequency is varying during this period). Because this is a type 2 loop, the error settles to zero as long as the jammer remains at the set-on frequency. When the jammer starts its pull-off at $t = 0.5$ s, the VCO follows with no cycle slipping. The error jumps to

$$(\phi_\epsilon)_{ss} = \frac{\dot{\omega}_d}{K} = \frac{5000}{10^4} = 0.5 \text{ rad}$$

The simulation model would allow ϕ_j and ϕ_v to continue indefinitely counting cycles, continuing the parabolic plots of Figure 5.12. In Figure 5.13, conditions were identical to Figure 5.12, except that the pull-off was begun immediately, without allowing time for the VCO to acquire the jammer frequency; acquisition had to be achieved on a signal that was pulling away in frequency. This required 1.5 s.[15]

A number of simulation runs were made, similar to the two discussed, with successively decreasing J/S values. The velocity tracker followed the false target's pull-off maneuver until J/S reached 3 dB. At this level the loop was on the verge of cycle slipping. A little additional noise or clutter would have induced slipping.[16]

[14]The presence of the target, 40 dB, below the jammer level has little or no effect on the transient behavior. The target simply determines the initial VCO frequency before the jammer turns on.

[15]In this plot and in Figure 5.14, automatic scaling of the plot was permitted in order to allow small details, such as final behavior of ϕ_ϵ, to be displayed. This resulted in different scales for ϕ_j and ϕ_v, causing the plots to cross over. They would not cross if plotted on the same scale. ϕ_v would always lag ϕ_j by a constant value after lock-on.

[16]Slipping of an occasional cycle does not make the VGPO less effective, but it does indicate that failure would occur should J/S fall a bit lower.

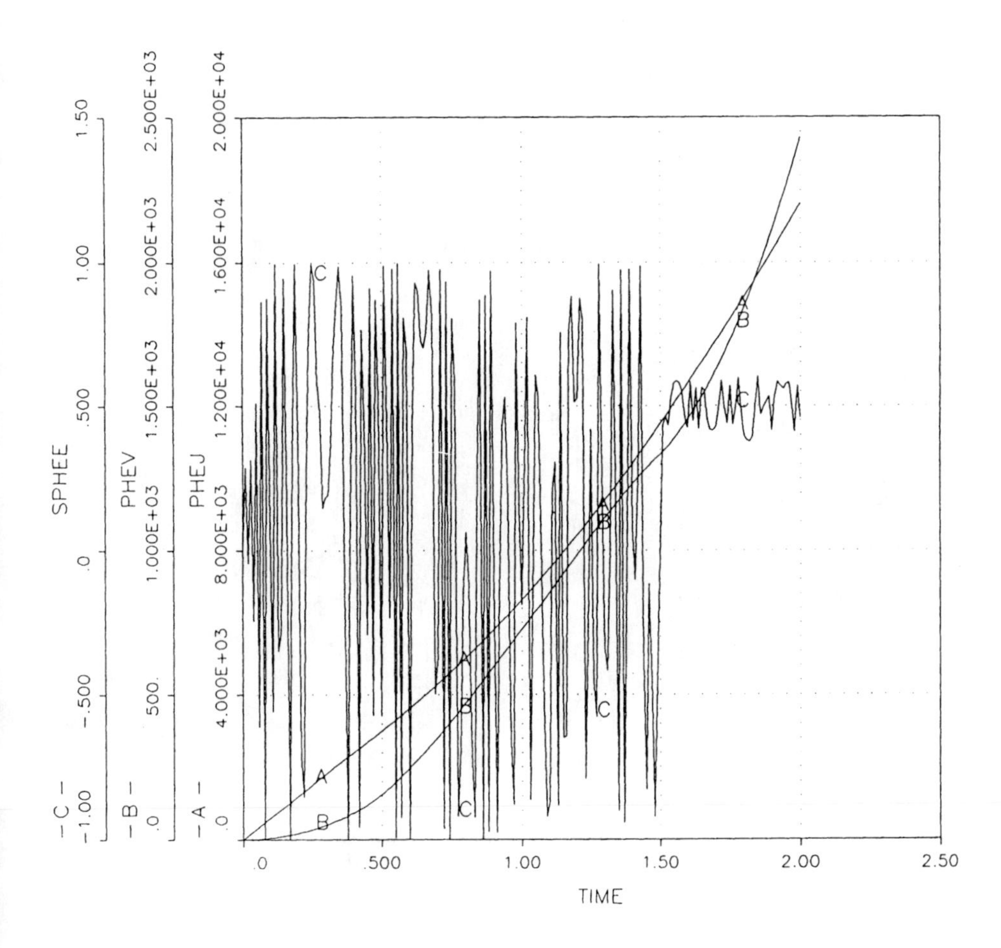

Figure 5.13 VGPO simulation with larger set-on error.

Figure 5.14 illustrates complete failure. In this case J/S was reduced to -3 dB. Now it is clear that the VCO remains locked to the target. Although there is considerable jitter in ϕ_v, its value never reaches 2π (no cycles are slipped). Recall that both ϕ_j and ϕ_v are measured relative to target phase.

Figure 5.14 is indicative of the accuracy with which the loop would follow a jammer with $J/S = +3$ dB. In Figure 5.14 it is following the signal with $S/J = +3$ dB.

For the sake of realism, the jammer should not abruptly make the transition from a constant doppler to a linearly changing doppler. Moreover, to make the deception as realistic as possible, range gate pull-off should take place at the same time, with $\dot{R}$ matched to ω_d.

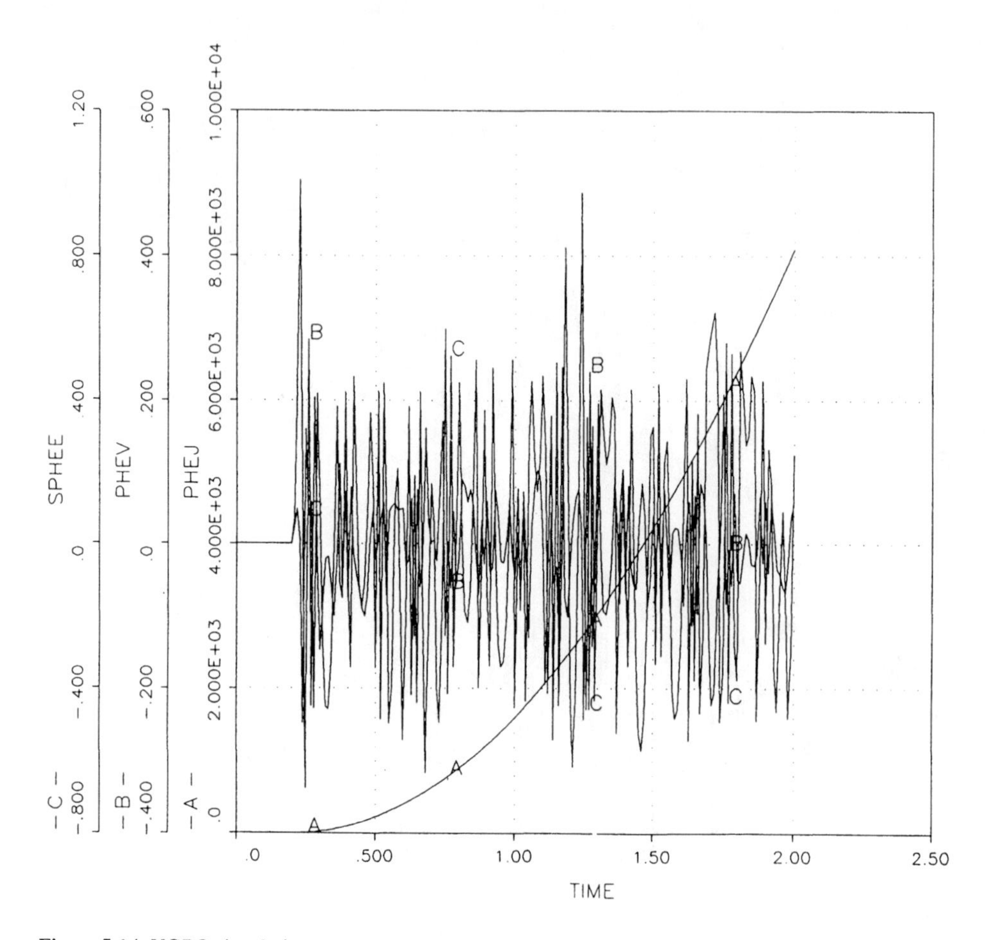

Figure 5.14 VGPO simulation at very low *J/S*.

REFERENCES

1. Skolnik, M.I., ed., *Radar Handbook,* 2nd Ed., McGraw-Hill, New York, 1990.
2. Morris, G.V., *Airborne Pulsed Doppler Radar,* Artech House, Norwood, MA, 1988.

Chapter 6
ON-OFF JAMMING OF AUTOMATIC GAIN CONTROL

6.1 BACKGROUND

A familiar application of AGC is in AM radio receivers, where it is referred to as *automatic volume control* (AVC). In that application the gain of the IF amplifier is controlled by an AVC voltage that is proportional to the average carrier level into the second detector. The AVC action keeps the carrier out of the IF amplifier at a nearly constant level in spite of large variations in received signal strength. The result is that, for a given degree of amplitude modulation, the audio level to the listener remains constant.

AGC for radar differs little in principle from AVC. Although AGC is applied to maintain a proper video level for displays (e.g., in ground mapping), our concern in this chapter is with the application of AGC to tracking radars. In this application the AGC operates on a range-gated signal with the gate centered on the echo from the target to be tracked. Without AGC, this echo signal amplitude would vary drastically with varying target cross section, target scintillation, and target range. The error detector of a coordinate tracker (range, angle, or doppler tracker) generally develops an error voltage that is proportional to both the coordinate error and the amplitude of the signal into the error detector. A tracking servo (see Appendix A) is designed to produce an error correction (e.g., a correction voltage or a correction torque) that is proportional to the error. The proportionality constant should be fixed at a specific design value, for it is a factor in the servo loop gain; it would be intolerable to have this factor vary with signal amplitude. It is the function of the AGC to maintain a constant average signal level into the trackers. As we shall see, the AGC forces the gain of the IF amplifier to be inversely proportional to the average level of the received signal. The AGC loop filter (a low-pass filter) performs the averaging, generally over many received pulses. Usually, the AGC detector incorporates a boxcar circuit (see Appendix B) that stretches the detected target pulse to a duration nearly equal to the PRI. Without stretching, the average of the pulse samples would be proportional to the transmit duty factor (pulsewidth ÷ PRI). Stretching increases the average, and consequently the AGC loop gain.

In Figure 6.1(a), v_1 and v_2 are the envelope amplitudes of the target return signal at the input and output of the gain-controlled IF amplifier. The amplifier voltage gain A is a function of control voltage v_c:

$$A = A(v_c)$$

The AGC detector develops a video pulse of amplitude v_2 and stretches it in a box-car circuit. Thus, for a steady target amplitude, the AGC detector develops a dc voltage equal to v_2. This voltage enters the voltage comparator along with dc reference voltage v_r. The comparator output, v_3, is given by

$$v_3 = \begin{cases} v_2 - v_r & \text{for } v_2 \geq v_r \\ 0 & \text{for } v_2 < v_r \end{cases}$$

This configuration is commonly called *delayed AGC*. The meaning of "delay" in this connection is not a time delay, but a withholding of control until v_2 exceeds reference level v_r. For very low level input signals, the IF amplifier operates at maximum gain. Once v_2 exceeds v_r, control voltage v_c begins to rise, causing the amplifier gain to be reduced. Not all AGC circuits incorporate the delay feature. If v_r is absent, gain control begins as soon as v_2 reaches a level high enough to develop a voltage out of the AGC detector. The detector diode itself has an inherent delay; it requires a finite input (a fraction of a volt) before any output is developed. Figure 6.1(a) shows a dc amplifier in the AGC loop. The same loop gain can be equally well provided by an IF amplifier ahead of the AGC detector. With high loop gain, the AGC action can be made so effective that large changes in v_1 produce only very small changes in v_2, with the result that the average level of v_2 remains close to reference voltage v_r.

The reader should understand that the output leveling function which we have described is operative only for input changes slow enough for the resultant changes in v_2 to be passed by the loop filter. More rapid signal amplitude fluctuations pass through the IF amplifier unaffected by AGC. We shall be particularly concerned with AGC for angle trackers employing sequential lobing (see Chapter 4)—for example, conical scan trackers and TWS radars. In these radars, target angle information must be extracted from target-induced amplitude modulation of the received signal. The AGC is required to maintain the average signal level into the tracker constant, but AGC action must not be so fast as to suppress the signal amplitude fluctuations produced by the scanning of the antenna beam across the target. The AGC loop filter must therefore have a low-pass response, passing slow changes in average signal level but not passing scan-frequency modulation com-

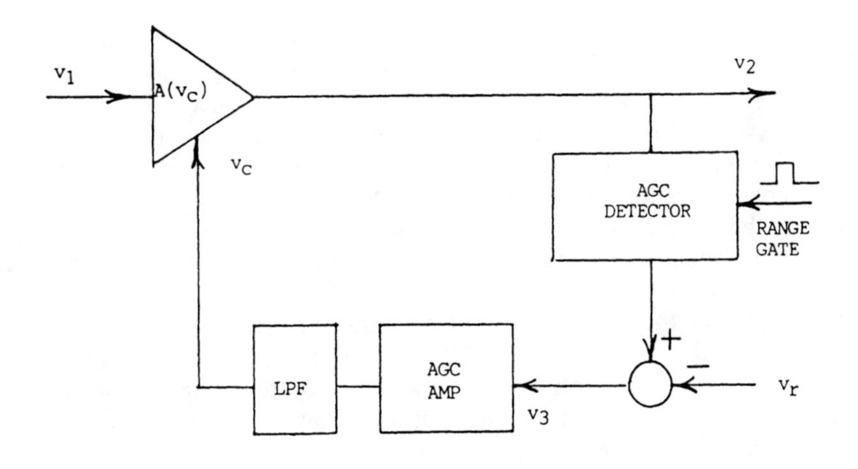

(a) AGC Loop Block Diagram

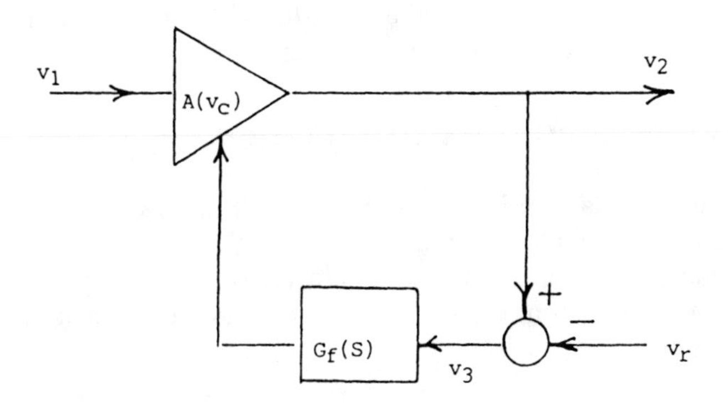

(b) Simplified Representation

Figure 6.1 AGC loop.

ponents. Such an AGC is called *slow AGC.* In Sections 6.2 and 6.3 we study the normal behavior of slow AGC in preparation for investigating, in Section 6.4, on-off jamming intended to disable such an AGC. Although the analysis that follows is concerned with slow AGC, we discuss briefly the possibility of using fast AGC in Sections 6.4 and 6.5, and explain that with fast AGC the target modulation waveform can be extracted from the AGC control voltage. This is necessary because fast AGC suppresses amplitude modulation in the output of the gain-controlled amplifier. With fast AGC, the fast AGC may prevent amplifier saturation entirely (provided the jamming is not beyond the AGC control range). If saturation does occur, recovery will be rapid because of the shorter time constant of the loop filter. Figure 6.15 shows the modulation available on the control voltage of a simulated fast AGC.

In Section 6.2 the AGC static control characteristics are developed. We find that for two very different control laws (linear and logarithmic) the feedback loop with high gain produces essentially the same static control characteristic. The logarithmic control law is closer to reality, but the linear law provides the means for analytically treating the effects of on-off jamming in Section 6.4.

In Section 6.3 we employ a small-signal linear model to study the passage of AM components through the gain-controlled amplifier. We find that AGC action forces the mean output carrier level to a value very close to the reference level (v_r of Figure 6.1), and that the AM index or degree of modulation is preserved.

In Section 6.4 we find that on-off jamming can disable the slow AGC by driving it from one extremity of the control characteristic to the other. Simulation results verify that the response of the AGC to such jamming conforms to the pattern predicted by analysis of the linear model.

6.2 STATIC CONTROL CHARACTERISTICS

The static control characteristic is the relation of output voltage to input voltage of the gain-controlled amplifier for slowly changing inputs (slow compared with the AGC response). Figure 6.2 is a plot of gain versus control voltage for two control laws, linear and logarithmic. The linear law provides a means for analyzing small-signal amplification in Section 6.3.

Linear Control Law

The linear control law depicted in Figure 6.2 is of the form:

$$A = A_0 - \mu v_c$$

Returning to the simplified AGC diagram of Figure 6.1(b), we have identified the transfer function (see Appendix B) of the loop filter as $G_f(S)$. We therefore let $G_f(0)$ represent the dc (low-frequency) transfer function of the filter. Then, for slowly changing inputs, v_c is related to v_2 by

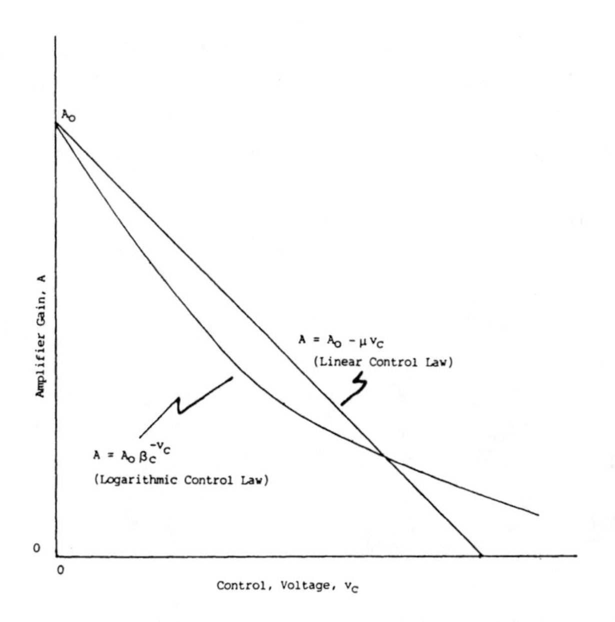

Figure 6.2 AGC, gain *versus* control voltage.

$$v_c = v_3 G_f(0) = \begin{cases} (v_2 - v_r)G_f(0) & \text{for } v_2 \geq v_r \\ 0 & \text{for } v_2 < v_r \end{cases}$$

For inputs below the control level ($v_1 < v_r/G_0$) the amplifier gain is G_0. For inputs within the control range:

$$A = A_0 - \mu(v_2 - v_r)G_f(0) \tag{6.1}$$

We shall designate the low-frequency content (average value) of the input and output as $\overline{v}_1$ and $\overline{v}_2$, respectively. The static control characteristic is then given by

$$\overline{A} = \frac{\overline{v}_2}{\overline{v}_1} = \frac{A_0 + \mu v_r G_f(0)}{1 + \mu \overline{v}_1 G_f(0)} \tag{6.2}$$

In the following example we plot $\overline{A}$ and $\overline{v}_2$ *versus* $\overline{v}_1$ for a specific design. In this example, the gain-controlled IF amplifier follows a preamplifier (with a gain of perhaps 30 to 40 dB), which has raised the minimum usable target return to a level of 10^{-4} V. The AGC control range extends from this input level upward for 67 dB. The AGC is designed to produce a standard output of 1 V; this output would, for example, feed a conical scan tracker. The instantaneous output dynamic range is from zero up to the 2.5-V limit level. This permits faithful reproduction of conical scan modulation patterns even for large tracking errors (modulation indexes approaching 100%).

Example 1: The parameters for this example are the following:

$$A_0 = 10^4;$$
$$\mu = 10^5\ \text{V}^{-1};$$
$$G_f(0) = 1.0;$$
$$v_r = 1.0\ \text{V};$$
$$A_{min} = 5.0 \text{ or } 14\ \text{dB};$$
$$V_{lim} = 2.5\ \text{V}.$$

A_{min} defines the upper end of the control range. When v_c reaches v_{cmax}, which causes A to be reduced to A_{min}, there is no further increase in v_c, so A remains at A_{min} for further increases in v_1. From (6.1) we find that

$$v_{cmax} = \frac{A_0 - A_{min}}{\mu} = 0.09995\ \text{V}$$

The value of v_{lim} is the limiting value of v_2 (no increase in v_2 results for further increases in v_1, once v_2 reaches v_{lim}). These limits on v_c and v_2 apply to both static and dynamic inputs. Figure 6.3 (the solid curves) describes the static control characteristic for this example. Control begins at $v_1 = 10^{-4}$ V and extends up to $v_1 \approx 0.22$ V, at which point $\overline{A}$ reaches $\overline{A}_{min}$. Thus, the input dynamic range is

$$20 \log \left(\frac{0.22}{10^{-4}} \right) = 67\ \text{dB}$$

Over this range the output rises from 1.0 to 1.1 V, a range of only 0.8 dB.

Logarithmic Control Law

Hughes [1, Chapter 1] states that most variable-gain amplifiers respond to control with an approximately linear relation between gain, expressed in dB, and control voltage. For this control law we write

$$A_{dB} = A_{0dB} - \beta v_c \tag{6.3}$$

with the substitution

$$\beta = 20 \log(\beta_c)$$

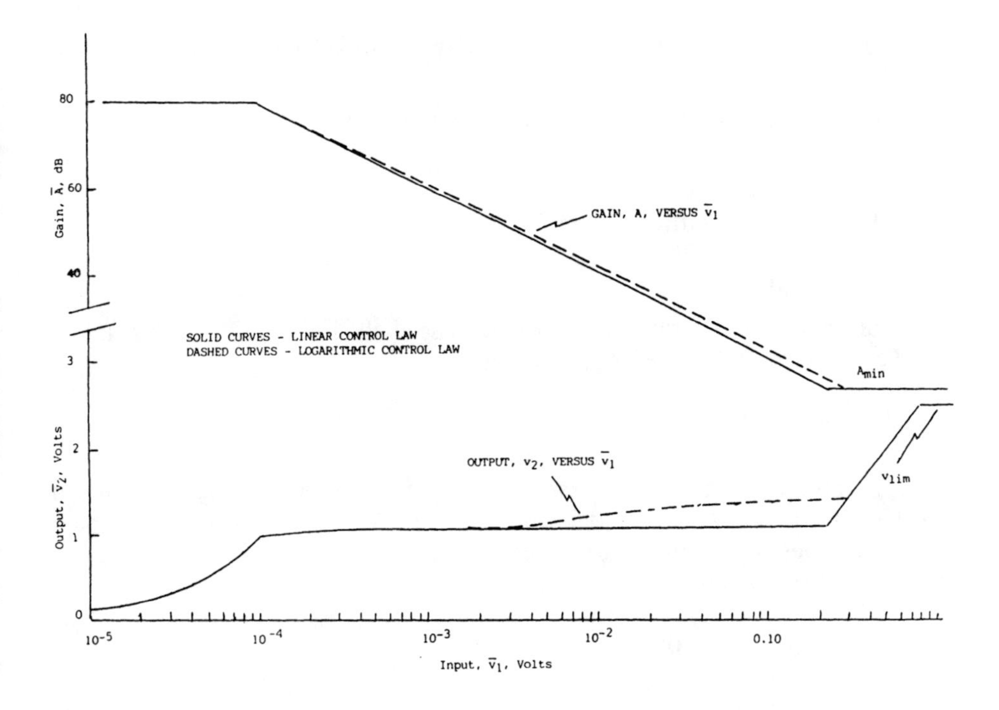

Figure 6.3 AGC static control characteristics.

Equation (6.3) can be rewritten as

$$A = A_0 \beta_c^{-v_c} \tag{6.4}$$

where $A_{\text{dB}} = 20 \log(A)$, and $A_{0\text{dB}} = 20 \log(A_0)$. This is the logarithmic control law plotted in Figure 6.2. We assume that, as in Figure 6.1(b), within the control range:

$$V_c(s) = [V_2(s) - v_r]G_f(s)$$

so, for the static case ($s = 0$) we can write

$$\bar{v}_c = (\bar{v}_2 - v_r)G_f(0)$$

The static control characteristic can be written as

$$\bar{A} = \frac{\bar{v}_2}{\bar{v}_1} = A_0 \beta_c^{(v_r - \bar{v}_2)G_f(0)}$$

In the following example we plot the static control characteristics of the AGC for a specific set of parameters.

Example 2: The parameters for this example are the following:

$$A_0 = 10^4;$$
$$v_r = 1.0 \text{ V};$$
$$A_{\min} = 5.0 \text{ or } 14 \text{ dB};$$
$$v_{\lim} = 2.5 \text{ V};$$
$$G_f(0) = 1.0.$$

This AGC is designed to serve the same purpose as the linear AGC of Example 1.

Equation (6.3), for $A = A_{\min}$, becomes

$$A_{\min\text{dB}} = 80 \text{ dB} - \beta v_{c\max} = 14 \text{ dB}$$

Let us, in this case, permit a 3-dB rise in output over the control range. This means that the output will rise from 1.0 V, for $v_1 = 10^{-4}$ V, to about 1.4 V at the point where A reaches $A_{\min}$. Thus, at the upper end of the control range, $v_c = v_{c\max}$ will have reached the value:

$$v_{c\max} = 1.4 - v_r = 0.4 \text{ V}$$

so from the equation for $A_{\min}$ we get

$$14 \text{ dB} = 80 \text{ dB} - 0.4\beta$$

We thus determine β to be

$$\beta = \frac{80 - 14}{0.4} = \frac{66}{0.4} = 165 \text{ dB/V}$$

For every 6-mV change in v_c there is 1-dB change in gain. The corresponding value of β_c is

$$\beta_c = 10^{165/20} = 10^{8.25}$$

The upper end of the control range is

$$v_1 = \frac{1.4}{5} = 0.28 \text{ V}$$

so the value of the control range is

$$20 \log \left(\frac{0.28}{10^{-4}} \right) = 69 \text{ dB}$$

The static control characteristic for this example,

$$\overline{A} = \frac{\overline{v}_2}{\overline{v}_1} = 10^4 (10^{8.25})^{v_r - \overline{v}_2} = 10^{12.25 - 8.25\overline{v}_2}$$

is plotted as the dashed curve in Figure 6.3.

The parameters for Examples 1 and 2 were chosen to provide essentially the same static control characteristics for the linear and logarithmic control laws. The two outputs diverge by only 2 dB over the entire 67-dB input range.

Although the high gain in the AGC loop forces the two static control characteristics to be very similar, the dynamic responses to low-level modulation are quite different. The gain for such modulation components is shown in Section 6.3 to depend on the slope of the control curve at the operating point. Although this slope is constant for the linear control law, it varies drastically over the range $0 < v_c < v_{c\max}$ for the logarithmic control law, as is evident in Figure 6.2.

6.3 SMALL-SIGNAL LINEAR MODEL

In this section we analyze the transmission of an amplitude-modulated signal through the gain-controlled amplifier. In the next section we find that on-off jamming is capable of taking over the AGC and preventing the passage of the signal.

In Figure 6.4 the curve labeled $A = A(v_c)$ describes the nonlinear relation of voltage gain to control voltage. Linear circuit analysis is inapplicable to the general study of the AGC loop. However, the amplification of small input modulation envelopes can be dealt with by the following linear model suggested by Maksimov [2, Chapter 5].

The AGC loop is that of Figure 6.1(b). We shall represent the input signal by

$$v_1 = \overline{v}_1 + \Delta v_1 \tag{6.5}$$

where $\overline{v}_1$ is the mean input voltage and Δv_1 is a small modulation about the mean:

$$\Delta v_1 \ll \overline{v}_1$$

The "mean" is simply defined to be an average over a time that is long relative to the response time of the AGC loop. We shall assume that $\overline{v}_1$ is large enough to produce an output $\overline{v}_2$ that exceeds v_r. The mean input to the loop filter is then

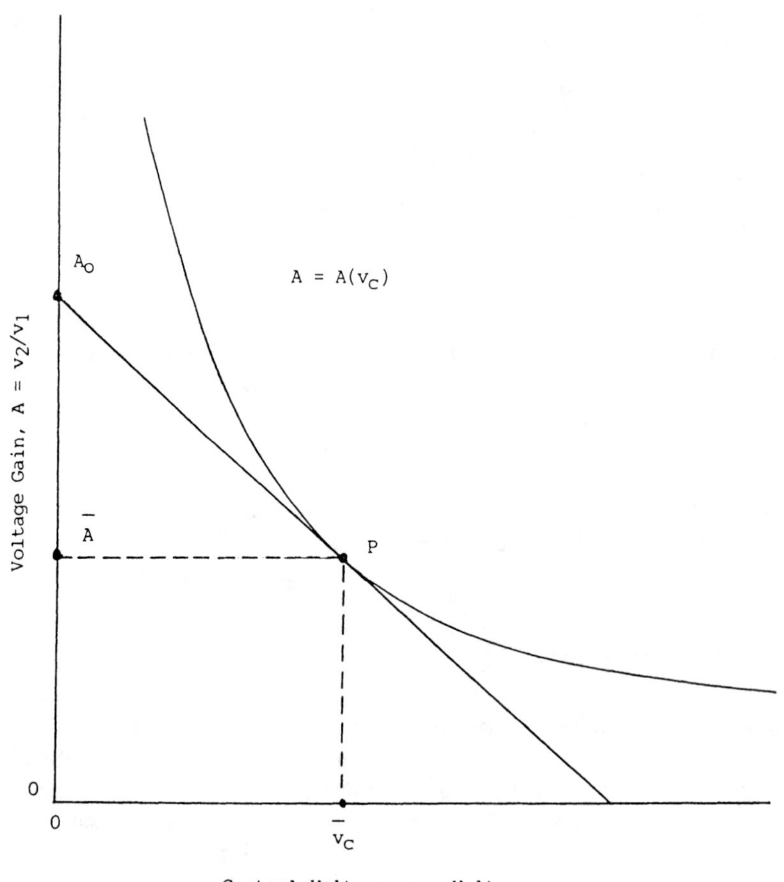

Figure 6.4 AGC control characteristics.

$$\overline{v}_3 = \overline{v}_2 - v_r$$

The mean control voltage is given by

$$\overline{v}_c = \overline{v}_3 G_f(0) = (\overline{v}_2 - v_r)G_f(0)$$

where $G_f(0)$ is the low-frequency (dc) value of the loop-filter transfer function. The operating point *P* on the curve of Figure 6.4 is therefore determined by

$$\overline{v}_2 = \overline{v}_1 \overline{A}(\overline{v}_c) \tag{6.6}$$

Although we cannot solve (6.6) explicitly for $\bar{v}_2$, we can readily determine P numerically, given $\bar{v}_1$ and the graph of $A(v_c)$ *versus* v_c. The behavior of the loop for small modulations about P is determined by the slope of the control curve in the vicinity of P. To this end, we replace the control curve by a straight line through point P, matching the slope of the control curve. The linear approximation is

$$A = A(v_c) = A_0 - \mu v_c \tag{6.7}$$

We then write

$$A = \bar{A} + \Delta A \tag{6.8}$$

where $\bar{A}$ is the value of A at P, and ΔA represents the variation in A that results from input voltage modulation. Equation (6.8) can be expressed as

$$A = \bar{A} + \frac{\partial A}{\partial v_c} \Delta v_c = \bar{A} - \mu \, \Delta v_c \tag{6.9}$$

We then replace the equation

$$v_2 = A v_1$$

by the following equation:

$$\bar{v}_2 + \Delta v_2 = (\bar{A} - \mu \, \Delta v_c)(\bar{v}_1 + \Delta v_1) \tag{6.10}$$

We obtain the relation describing the operating point by setting $\Delta v_1 = \Delta v_2 = 0$:

$$\bar{v}_2 = \bar{A}\bar{v}_1 = (A_0 - \mu\bar{v}_c)\bar{v}_1 = [A_0 - \mu G_f(0)(\bar{v}_2 - v_r)]\bar{v}_1$$

so the static (low-frequency) gain, $\bar{A}$, is

$$\bar{A} = \frac{\bar{v}_2}{\bar{v}_1} = \frac{A_0 + \mu G_f(0) v_r}{1 + \mu G_f(0)\bar{v}_1} \tag{6.11}$$

As would be expected, (6.11) is precisely the static gain, (6.2), derived on the basis of an assumed linear control law. As (6.11) indicates, the static gain decreases as $\bar{v}_1$ increases, thereby suppressing the effects of variation of mean input level on the output level. The relationship between input and output modulation components extracted from (6.10) is

$$\Delta v_2 = \bar{A} \, \Delta v_1 - \mu\bar{v}_1 \, \Delta v_c - \mu \, \Delta v_c \, \Delta v_1 \tag{6.12}$$

Let us discard the second-order term, $\mu \; \Delta v_c \; \Delta v_1$, and make the following change in notation for the modulation voltage components: replace Δv_1 by e_1, Δv_2 by e_2, and Δv_c by e_c. We wish to derive a transfer function relating output to input, so $E_1(s)$, $E_2(s)$, and $E_c(s)$ will represent the Laplace transforms of e_1, e_2, and e_c. $E_c(s)$ is related to $E_2(s)$ through the transfer function, $G_f(s)$, of the loop filter:

$$E_c(s) = E_2(s)G_f(s)$$

From (6.12) the equation relating modulation components is

$$e_2 = \overline{A}e_1 - \mu\overline{v}_1 e_c$$

The transform of this equation is

$$E_2(s) = \overline{A}E_1(s) - \mu\overline{v}_1 G_f(s)E_2(s)$$

which results in the transfer function:

$$\frac{E_2(s)}{E_1(s)} = H(s) = \overline{A}\,\frac{1}{1 + \mu\overline{v}_1 G_f(s)} \qquad (6.13)$$

Let us consider a signal arising from conical scan modulation of a target return:

$$v_1 = \overline{v}_1[1 + m\cos(\omega_s t)]$$

where $m \ll 1.0$ is the modulation index and ω_s is the radian scan frequency. The value of m is proportional to the angular-tracking error [3, Chapter 8], so the tracking servo must receive a signal that preserves the value of m but is unaffected by changes in input level $\overline{v}_1$. We shall see that the AGC performs this function. In accordance with our notation adopted for small modulations:

$$e_1 = \overline{v}_1 m\cos(\omega_s t)$$

and

$$E_1(j\omega_s) = E_1(s)|_{s=j\omega_s} = \overline{v}_1 m$$

We obtain the modulation amplitude at the output of the gain-controlled amplifier from (6.13) with s replaced by $j\omega_s$:

$$E_2(j\omega_s) = H(j\omega_s)E_1(j\omega_s) = \overline{A}\overline{v}_1 \left[\frac{m}{1 + \mu\overline{v}_1 G_f(j\omega_s)} \right]$$

The first factor, $\overline{A}\overline{v}_1$, is just the mean output, $\overline{v}_2$, given by

$$\overline{v}_2 = \overline{A}\overline{v}_1 = \frac{A_0 + \mu G_f(0)v_r}{1 + \mu G_f(0)\overline{v}_1}\overline{v}_1$$

For $\mu G_f(0)\overline{v}_1 \gg 1.0$, this equation becomes

$$\overline{v}_2 = \frac{A_0 + \mu G_f(0)v_r}{\mu G_f(0)}$$

This factor is independent of input amplitude $\overline{v}_1$ (as we already knew from our study of static control characteristics), so the output modulation amplitude is

$$E_2(j\omega_s) = \overline{v}_2 m \left[\frac{1}{1 + \mu\overline{v}_1 G_f(j\omega_s)} \right]$$

The frequency-dependent factor depends on amplitude $\overline{v}_1$, but if the low-pass filter is designed so that ω_s is well beyond the cut-off frequency:

$$\mu\overline{v}_1 G_f(j\omega_s) \ll 1.0$$

and the output becomes

$$E_2(j\omega_s) = \overline{v}_2 m \tag{6.14}$$

This is the desired result, for, as we have shown, with high AGC loop gain $\overline{v}_2$ becomes essentially constant (nearly equal to v_r) regardless of input amplitude $\overline{v}_1$. The modulation waveform into the angle tracker is

$$e_2(t) = \overline{v}_2 m \cos(\omega_s t)$$

resulting in a tracking servo loop gain that is constant and independent of $\overline{v}_1$.

To illustrate the requirement on the AGC loop filter, let us assume a single-pole RC filter:

$$G_f(j\omega) = \frac{1}{j\omega T_1 + 1}$$

where $T_1 = RC$. The modulation transfer function obtained from (6.13), with $s = j\omega$, becomes

$$H(j\omega) = \frac{E_2(j\omega)}{E_1(j\omega)} = \frac{\bar{A}}{1 + \mu\bar{v}_1}\left[\frac{j\omega T_1 + 1}{j\omega T_2 + 1}\right]$$

where

$$T_2 = \frac{T_1}{1 + \mu\bar{v}_1}$$

Figure 6.5 is a plot of $\log|H(j\omega)|$ (or $|H(j\omega)|$ in dB) *versus* $\log \omega$. At low frequencies (ωT_1 and $\omega T_2 \ll 1.0$) the transfer function is

$$H(j\omega) = \frac{\bar{A}}{1 + \mu\bar{v}_1}$$

In the low-frequency region, the modulation components are attenuated by the factor $1/(1 + \mu\bar{v}_1)$ because the modulation components are passed by the loop filter. In the high-frequency region (ωT_1 and $\omega T_2 \gg 1.0$) the transfer function is

$$H(j\omega) = \frac{\bar{A}}{1 + \mu\bar{v}_1}\frac{T_1}{T_2} = \bar{A}$$

The filter time constant T_1 would be chosen to place ω_s in the high-frequency region, so the output modulation amplitude becomes

$$E_2(j\omega_s) = H(j\omega_s)E_1(j\omega_s) = \bar{A}\bar{v}_1 m = \bar{v}_2 m$$

as we found in (6.14).

Note that AGC action makes the filter corner frequency, ω_2, depend on input signal amplitude:

$$\omega_2 = \frac{2\pi}{T_2} = \frac{2\pi}{T_1}(1 + \mu\bar{v}_1) = \omega_1(1 + \mu\bar{v}_1)$$

This means that $T_1 = RC$ must be chosen large enough to ensure that $\omega_2 < \omega_s$ for the largest anticipated value of $\bar{v}_1$ (e.g., the value of $\bar{v}_1$ corresponding to the upper end of the AGC control range).

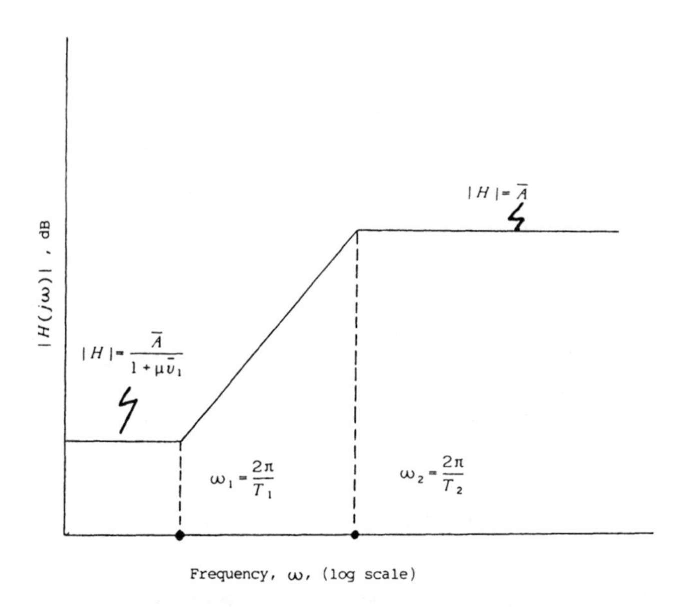

Figure 6.5 Frequency dependence of AGC modulation transfer function, $H(J\omega)$.

6.4 THE EFFECT OF ON-OFF JAMMING

In the preceding sections we have seen that a properly functioning AGC normalizes the mean level of the received signal and preserves the amplitude modulation index. In this section we demonstrate that on-off jamming can disrupt this AGC function.

On-off jamming that attacks the AGC function of a radar receiver is sometimes called *AGC deception* because it forces the AGC to adjust gain on the basis of jamming bursts rather than on the target return signal. Figure 6.6 shows jamming bursts of duration T_{on} followed by quiet periods of duration T_{off}. The envelope of the signal is seen at amplitude v_s, well below level v_j of the jamming. As in the preceding sections, this diagram shows not individual pulses but signal and jamming envelopes as they would appear after detection and processing by a boxcar circuit. The jammer burst actually consists of a group of transponder or repeater pulses timed to coincide with target return pulses (see Chapter 3).

The objective of this on-off jamming is to cause the gain-controlled amplifier to cycle between the following two states:

- During T_{on} the amplifier output is at limit level v_{lim} because of the large jammer input. All signal modulation is lost as a result of output limiting.

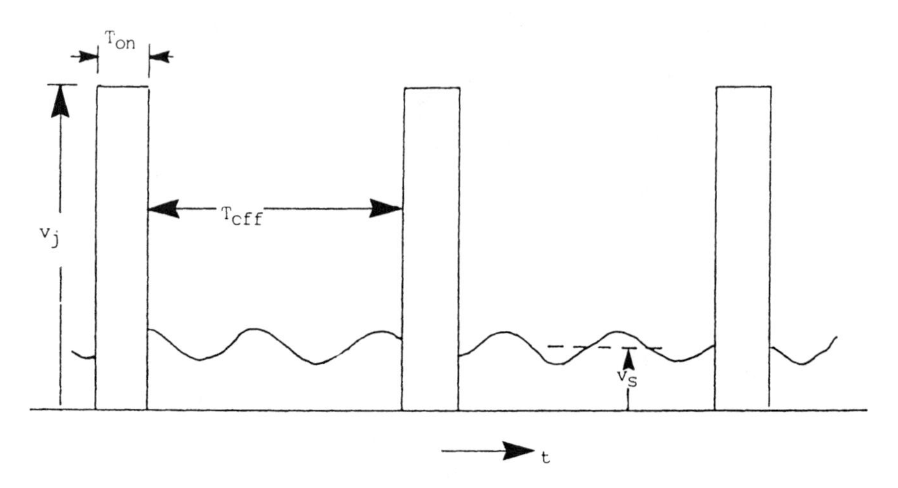

Figure 6.6 On-off jamming plus signal at receiver input.

- During T_{off} the amplifier gain is so low (because of the large AGC filter capacitor voltage developed during T_{on}) that the output signal is too small to be of any use.

If the jammer succeeds, the radar is left entirely without a usable target return signal. Clearly, to succeed, the jammer parameters must satisfy the following requirements:

1. The received jammer signal must be large enough to produce limiting.
2. T_{on} must be long enough to charge the AGC filter capacitor to a high enough voltage to produce the desired low value of gain at jammer turnoff.
3. T_{off} must not be so long as to permit the AGC filter to discharge to a degree that permits recovery to a useful value of gain.

Actually, neither requirement 2 nor 3 is absolute. The value of T_{on} can be greater than the time required for the gain-controlled amplifier to recover from limiting. The gain upon recovery from limiting will be low, but if the jammer remains on somewhat longer the gain will be reduced even further. The lower the gain at jammer turnoff, the longer will be the time required for the amplifier to return to useful amplification of the signal. Moreover, T_{off} can be longer than the time required to reach a usable level of signal output. The gain recovers as an exponential function of time, so the amplifier output v_2 recovers in the same manner. The tracking servo that utilizes v_2 performs an averaging of this signal, so it is the average value v_2 (averaged over a jamming cycle) that determines the effectiveness of the jamming. The track servos that depend on v_2 are designed to operate with a level of v_2 very nearly equal to reference level v_r. If the average of v_2 is appreciably below v_r, the servo loop gain will be below the design value by a corresponding degree, and tracking will be degraded or even disabled. We shall later measure jamming effectiveness in terms of the degree to which the average output signal is suppressed below v_r.

AGC Transient Analysis

On-off jamming drives the AGC over its entire control range and beyond. We cannot expect the linear model of Section 6.3 that we used to study small-signal transmission to provide a quantitatively accurate description of AGC behavior under such violent excitation. Nevertheless, a linear model [2, Chapter 5] can provide insight into the various transient phases induced by the jamming. The transitions between transient phases correspond to the beginning and end of output limiting and the beginning and end of AGC control action (passage through the point at which $v_2 = v_r$ as v_2 is rising or falling). With the linear model each transient phase involves an exponential transient voltage waveform. The linear model results in correct values of voltages at the transition points, but the durations of the transient phases are not predicted accurately by the linear model because the actual AGC with its nonlinear control characteristic does not have a single exponential decay constant (or time constant) for each phase, but one that varies as the operating point moves along the nonlinear control curve (see Figure 6.4) as the transient progresses.

The AGC loop of Section 6.1 is repeated here as Figure 6.7, showing an RC filter with transfer function:

$$G_f(s) = \frac{1}{sT + 1}$$

where $T = RC$. For the linear control law, the following set of equations governs loop behavior:

$$v_2 = Av_1 \qquad \text{for } v_2 < v_{\text{lim}}$$

$$A = A_0 - \mu v_c$$

$$v_c = \frac{q}{c}$$

$$v_3 = R\dot{q} + v_c$$

$$v_3 = \begin{cases} v_2 - v_r & \text{for } v_2 \geq v_r \\ 0 & \text{for } v_2 < v_r \end{cases}$$

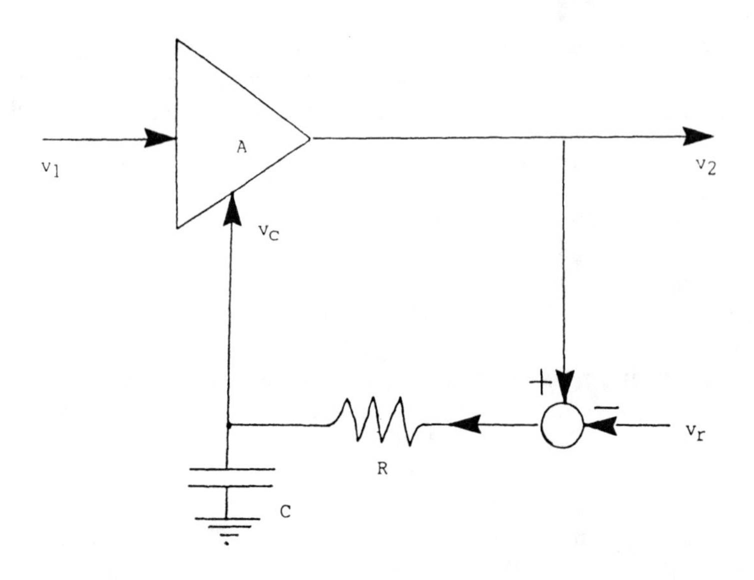

Figure 6.7 AGC diagram.

where q is the charge on the capacitor and $\dot{q}$ is the current (time derivative of charge) flowing through the resistor into the capacitor. We assume that the voltage comparator with output v_3 has negligible output impedance compared with R. As in Section 6.2, we assume output limiting when v_2 reaches v_{lim}.

The circuit of Figure 6.7 has the following transient regimes:

a. $v_2 < v_r$,
   $v_3 = 0$ (control inoperative);
b. $v_2 = v_{\text{lim}}$,
   $v_3 = v_{\text{lim}} - v_r$ (control inoperative);
c. $v_r < v_2 < v_{\text{lim}}$,
   $v_3 = v_3(t) = v_2(t) - v_r$ (control operative).

In regimes a and b the AGC loop can be said to be open; variations in v_c do not propagate around the loop to influence filter input v_3. The loop is closed in regime c. We can find the transient behavior of the loop by applying Laplace transformations to the five equations describing loop behavior. In the following discussion we take a shortcut to the transient solutions on the basis of our knowledge of the loop's static behavior (see Section 6.2) and the effect of feedback on the effective time constant of the loop (see Section 6.3).

The transient waveform of primary interest is the capacitor voltage v_c, because this voltage is continuous from one transient phase to the next. The initial value of v_c for a given phase is just the final value from the preceding phase. We use

$v_c(0+)$ to represent the initial capacitor voltage at the beginning of any transient phase. It is a simple matter to write the expressions for v_c for regimes a and b.

In regime a there is no output from the voltage comparator, so the capacitor simply discharges from its initial value through the resistor (and the negligible output resistance of the comparator):

$$v_c = v_c(0+)e^{-t/T} \tag{6.15}$$

In regime b the v_c transient has two terms. One represents the exponential decay of the initial charge as in (6.15). The other represents the exponential rise toward the voltage $v_{\text{lim}} - v_r$ that would eventually be reached if the amplifier were to remain in limit. The result is

$$v_c = v_c(0+)e^{-t/T} + (v_{\text{lim}} - v_r)[1 - e^{-t/T}] \tag{6.16}$$

In regime c the transient is more complex because feedback is operative. The feedback has two effects: it determines the ultimate voltage that v_c would reach if the loop could remain in this regime indefinitely (no output limiting and no cut-off of transmission through the comparator), and it determines the effective time constant of the loop.

We use $v_{c\infty}$ for the voltage that would eventually be reached and T' for the effective time constant. We again have two transient terms as in (6.16). The term representing the decay of initial charge differs from that of regimes a and b only in that T' replaced the time constant $T = RC$. The second term is similar to the second term of (6.16) but its time constant is also T'. The result is

$$v_c = v_c(0+)e^{-t/T'} + v_{c\infty}[1 - e^{-t/T'}] \tag{6.17}$$

The value of T' is precisely the value of the effective time constant T_2 that appears in Section 6.3:

$$T' = \frac{T}{1 + \mu\bar{v}_1} \tag{6.18}$$

Thus, for large values of $\mu\bar{v}_1$ we find the exponential transients to be much faster than in regimes a and b. The value of $v_{c\infty}$ is precisely the value that comes out of the static analysis of Section 6.2 or the analysis of Section 6.3, in which the linear control law is assumed. If the loop could remain in regime c indefinitely, the gain would reach the steady-state value of (6.1):

$$A_\infty = \overline{A} = \frac{A_0 + \mu v_r G_f(0)}{1 + \mu\bar{v}_1 G_f(0)}$$

Note that $G_f(0)$ is 1.0 and that $\bar{v}_1$ is simply equal to the constant[1] input voltage v_1 (either v_s or v_j of Figure 6.6). The ultimate output is

$$v_{2\infty} = v_1 \bar{A} = v_1 \frac{A_0 + \mu v_r}{1 + \mu v_1}$$

The ultimate capacitor voltage differs from $v_{2\infty}$ only by v_r:

$$v_{c\infty} = v_{2\infty} - v_r = \frac{A_0 v_1 - v_r}{1 + \mu v_1} \tag{6.19}$$

We add a subscript s or j to T' and $v_{c\infty}$ according to whether v_1 is equal to v_s or v_j. Therefore if the AGC is in regime c during T_{off}, we have

$$T' = T'_s = \frac{T}{1 + \mu v_s} \tag{6.20}$$

$$v_{c\infty} = v_{cs\infty} = \frac{A_0 v_s - v_r}{1 + \mu v_s} \tag{6.21}$$

If the AGC is in regime c during T_{on} we have

$$T' = T'_j = \frac{T}{1 + \mu v_j} \tag{6.22}$$

$$v_{c\infty} = v_{cj\infty} = \frac{A_0 v_j - v_r}{1 + \mu v_j} \tag{6.23}$$

The transition from regime b to regime c as the amplifier recovers from saturation occurs with

$$v_c = v_{c1} = \frac{A_0 - v_{\text{lim}}/v_j}{\mu} \tag{6.24}$$

The transition from regime a to regime c as v_2 reaches reference level v_r occurs with

$$v_c = v_{c3} = \frac{A_0 - v_r/v_s}{\mu} \tag{6.25}$$

[1]For this analysis we ignore any amplitude modulation of the signal envelope.

Figure 6.8 is a sketch of input voltage (v_1) and capacitor voltage (v_c) waveforms. We assume that the T_{on}-T_{off} jamming cycle has been repeated long enough for the v_c response to have settled to a truly periodic pattern. When the jammer is turned on, v_{c0} is the capacitor voltage. It is low, so the amplifier gain is high, thus resulting in output limiting (regime b). Thus, during interval T_1 the capacitor voltage obeys (6.16) with $v_c(0+) = v_{c0}$. When v_c reaches v_{c1}, the gain has been reduced sufficiently to terminate output limiting, so there is a transition to regime c and to (6.17) with $v_c(0+) = v_{c1}$ and $T' = T'_j$ and $v_{c\infty} = v_{cj\infty}$. This continues for an interval T_2 at the end of which the jammer turns off. Now, v_c has reached a rather high value, with v_{c2} causing amplifier gain A to be so low that, with v_s as the input, $v_2 < v_r$. Thus, the next transient phase is in regime a. During the interval of duration T_3 the capacitor voltage obeys (6.15) with $v_c(0+) = v_{c2}$. At the end of interval T_3 the capacitor voltage has reached v_{c3}, a value low enough to produce a gain A large enough to return to AGC control ($v_2 \geq v_r$). Thus the final transient phase is back in regime c. The capacitor voltage now obeys (6.17) with $v_c(0+) = v_{c3}$, $T' = T'_s$, and $v_{c\infty} = v_{cs\infty}$. This transient phase of duration T_4 ends when the jammer is turned on and the cycle is repeated. The four transient phases are labeled I through IV in Figure 6.8.

Given this model of the AGC of a victim radar, the jammer programmer would select a period, $T_{on} + T_{off}$, and a duty factor, $T_{on}/(T_{on} + T_{off})$, to disable the radar AGC function. Presumably, values of v_j and v_s would have been estimated for the scenario under consideration. Arriving at a suitable program would be a tedious process if it were done analytically, beginning when the jammer is first turned on and working through enough jamming cycles to arrive at a repetitive response. Access to an analog or digital simulator would simplify the process.

For illustrative purposes we can reverse the process, assuming values for the capacitor voltages at the transition points between transient phases (and assuming a repetitive response). Given v_s and v_j, the values of v_{c1} and v_{c3} are known ((6.24) and (6.25)), so it is only necessary to assume realizable values for v_{c0} and v_{c2}. These values are asymptotic to $v_{cs\infty}$ and $v_{cj\infty}$, respectively. It is necessary[2] that $v_{c0} < v_{c3}$ and that $v_{c2} > v_{c1}$.

For each assumed set of v_c values, we can compute the durations T_1, T_2, T_3, and T_4:

$$\frac{T_1}{T} = \ln\left(\frac{v_{\lim} - v_r - v_{c0}}{v_{\lim} - v_r - v_{c1}}\right) \tag{6.26}$$

$$\frac{T_2}{T} = \frac{\ln[1 - v_{c1}/v_{cj\infty})/f]}{1 + \mu v_j} \tag{6.27}$$

[2]If all four transient phases are to exist.

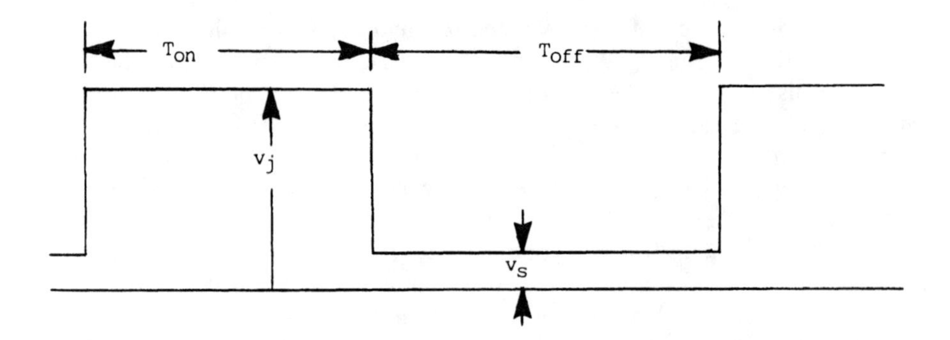

(a) Amplifier Input Envelope, v_1

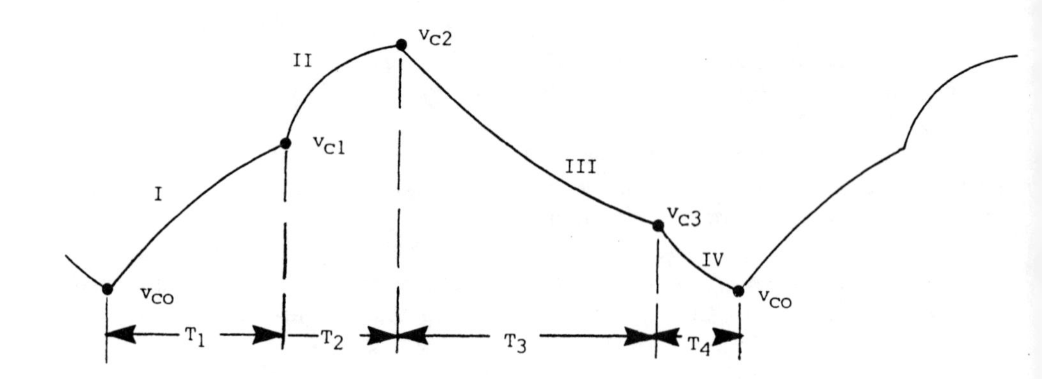

(b) Filter Capacitor Voltage, v_c

Figure 6.8 Input and filter capacitor waveforms.

$$\frac{T_3}{T} = \ln\left(\frac{v_{c2}}{v_{c0}}\right) \tag{6.28}$$

$$\frac{T_4}{T} = \frac{\ln\{(v_{c0}/v_{cs\infty} - 1)/g\}}{1 + \mu v_s} \tag{6.29}$$

In these equations, f and g are the fractional deviations from $v_{c\infty}$ that prevail when the values of v_{c2} and v_{c0} are reached; that is,

$$v_{c2} = (1 - f)v_{cj\infty} \tag{6.30}$$

$$v_{c0} = (1 + g)v_{cs\infty} \tag{6.31}$$

Once values of T_1, T_2, T_3, and T_4 have been computed, the jammer programmer must decide whether the resultant requirements on the cycle period and duty factor are acceptable. Presumably, a low value of duty factor is desired to minimize average power requirements and to leave free time for responding to other threats, but the main consideration is whether the trial parameters ensure that the average signal out of the gain-controlled amplifier is driven to a sufficiently low value to disable the radar tracker. Referring to Figure 6.8(b), we understand that no signal is available during phase I (output limiting on jamming). During phase II, both jamming and signal components are present at the amplifier output. If the jamming were truly of constant amplitude (no modulation to interfere with the signal) it is conceivable that a faint signal component could be observed in v_2. However, the amplifier gain in this phase is so low as to make this unlikely. Moreover, the speed of this transient is so great (T'_j so small) that the duration will be a very small fraction of the total jamming cycle. During phase III, the signal is present in the output, but, because the gain is recovering from a low value, the output tends to be low. Only during phase IV does the output signal rise to (and slightly above) reference level v_r. Again, this phase may have a short duration (T'_s will be small) if $\mu v_1 \gg 1.0$.

It is not difficult to compute an average value of signal output:

$$\bar{v}_{2s} = \frac{1}{P}\int_0^P v_{2s}\, \mathrm{d}t \tag{6.32}$$

where $v_{2s} = v_{2s}(t)$ is the exponential envelope of the signal at the amplifier output, and $P = T_1 + T_2 + T_3 + T_4$ is the cycle period. This envelope for any transient phase is simply

$$v_{2s}(t) = [A_0 - \mu v_c(t)]v_s \tag{6.33}$$

where $v_c(t)$ is the waveform of v_c for the transient phase under consideration.

Figure 6.9 is a result of the computation process just described,[3] using the linear AGC model with parameters identical to those of Example 1 in Section 6.2. The loop filter was the simple RC filter of Figure 6.7. The input signal level was 0.001 V, and the jamming level was 0.01 V (J/S = 20 dB). Equations (6.26)

[3]The computation included the output signal from phases II, III, and IV, but the phase II component always proved to be negligible.

through (6.28) yield period and duty factor values. We obtain signal-suppression values:

$$\text{signal suppressions} = \frac{\bar{v}_{2s}}{v_r}$$

from (6.32) and (6.33). The area of Figure 6.9 bounded by the curves labeled f = 0.1 and $f = 10^{-5}$ encompasses essentially the entire domain of realistic jamming parameters (period and duty factor). We derived these two bounding curves by fixing the parameter f of (6.30) and varying the parameter g of (6.31). For each computed coordinate point (period and duty factor) we determined the value of signal suppression. We also computed additional curves on the interior of the bounded area so that we could sketch the signal-suppression contours (signal-suppression values in dB are labeled on the contours).

The following discussion explains why Figure 6.9 encompasses the domain of realistic jamming. The right-hand boundary ($f = 10^{-5}$) allows transient phase II to carry v_{c2} so close to $v_{cj\infty}$ that nothing would be gained by allowing that phase to continue further; there would simply be an increase in duty factor (and jamming energy) with no payoff. The left-hand boundary ($f = 0.1$) brings v_{c2} down so close to v_{c1} that little of phase II remains. Thus, most of the very fast capacitor charging that prevails in phase II is sacrificed. Because v_{c2} has been reduced, the recovery in phase IV is shortened, hence, the shorter period than at the right-hand boundary for a given degree of signal suppression.

The values of g begin with $g = 10^{-6}$ at the top of the plot. With such a small value of g, transient phase IV is allowed to progress so far that it is the main contributor to $\bar{v}_{2s}$. Moving downward, g increases, reaching a value of 0.01 at about the level of the 6-dB contour. At this level of g, the duration of phase IV has been reduced to zero ($T_4 = 0$). Thus, signal-suppression values no greater than 6 dB are attainable as long as transient phase III is allowed to run to completion (v_c allowed to drop to v_{c3}) even when none of phase IV remains.

For greater suppression, we must truncate the phase III transient by initiating the next jammer burst. This is what occurs in the portion of the plot below the 6-dB contour. In that region we define a new parameter g' to describe the truncation process:

$$v_{c0} = (1 + g')v_{c3} \tag{6.34}$$

The greater the value of g', the shorter becomes the duration of the phase III transient and, consequently, the shorter is the jamming cycle period. The value of g' is 10^{-3} at the level of the 6-dB contour, and it reaches a value of 0.83 at the floor of the plot (the boundary of the shaded area) where signal suppression approaches 15 dB.

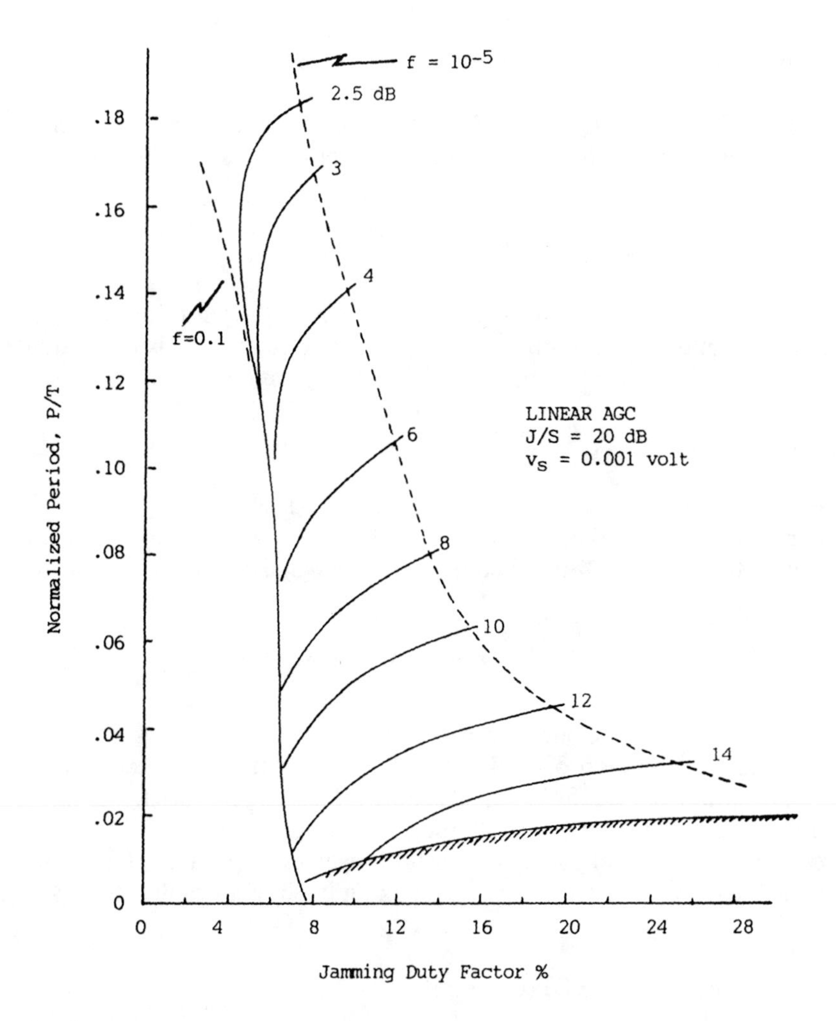

Figure 6.9 Signal suppression by on-off jamming.

The shaded floor region corresponds to jammer cycle periods so short that the loop filter passes little more than the dc component of the v_3 pulse produced by the jamming burst. The v_3 waveform is a train of pulses of amplitude

$$v_{3j} = v_{2j} - v_r$$

where v_{2j} is the amplitude of v_2 during the jamming burst (if the jamming produces output limiting, $v_{2j} = v_{\text{lim}}$). The dc component of the v_{3j} pulse train then becomes the dc component of v_c. Its value is

$$\bar{v}_c = \bar{v}_{3j} = (v_{2j} - v_r)\mathrm{DF} \tag{6.35}$$

where $\mathrm{DF} = T_{on}/(T_{on} + T_{off})$ is the duty factor of the jamming. With v_c remaining essentially constant at $\bar{v}_c$, the amplifier gain also remains essentially constant at the value:

$$A_{\mathrm{HF}} = A_0 - \mu\bar{v}_c = A_0 - \mu\mathrm{DF}(v_{2j} - v_r) \tag{6.36}$$

The subscript HF indicates that the cycle frequency of the jamming is so high that the ac components of v_3 are not passed by the loop filter. If v_j is high enough to produce output limiting, $v_{2j} = v_{\mathrm{lim}}$, and the gain expression becomes

$$A_{\mathrm{HF\,lim}} = A_0 - \mu(v_{\mathrm{lim}} - v_r)\mathrm{DF} \tag{6.37}$$

Under these conditions the signal output during T_{off} is $v_s A_{\mathrm{HF\,lim}}$.

For small values of the jamming duty factor, the average output signal is also nearly equal to that value multiplied by v_s, so the signal-suppression factor is

$$\text{signal suppression} = \frac{\bar{v}_{2s}}{v_r} = \frac{v_s}{v_r}[A_0 - \mu(v_{\mathrm{lim}} - v_r)\mathrm{DF}] \tag{6.38}$$

For the AGC parameters of our example, with an input signal of $v_s = 10^{-3}$ V, (6.38) yields a signal suppresssion of 0 dB when DF = 6%. If DF is increased to 1/15 or 6⅔%, that equation yields infinite signal suppression (the gain goes to zero). Clearly, this would not actually happen. As DF increases, the gain would drop until v_{2j} drops below v_{lim}, and (6.37) and (6.38) would no longer be applicable. The AGC loop would reach equilibrium with $v_{2j} < v_{\mathrm{lim}}$, making (6.36) applicable, so v_{2j} would be given by

$$v_{2j} = A_{\mathrm{HF}}v_j = [A_0 - \mu\mathrm{DF}(v_{2j} - v_r)]v_j$$

Solving this equation for $v_{2j}/v_j = A_{\mathrm{HF}}$ yields

$$A_{\mathrm{HF}} = \frac{v_{2j}}{v_j} = \frac{A_0 + \mu v_r \mathrm{DF}}{1 + \mu v_j \mathrm{DF}} \tag{6.39}$$

The signal suppression is then

$$\text{signal suppression} = \frac{v_s A_{\mathrm{HF}}}{v_r} = \frac{v_s(A_0 + \mu v_r \mathrm{DF})}{v_r(1 + \mu v_j \mathrm{DF})} \tag{6.40}$$

As (6.38) and (6.40) indicate, when the jamming cycle frequency becomes so high that the output transients are not passed by the filter, the signal-suppression factor becomes a function only of DF (for a given combination of v_s and v_j). This implies that the signal-suppression contours become vertical lines in the shaded floor region of the plot. The contours are shown turning downward into that region. Computations of signal suppression via (6.40) yield values compatible with the values of suppression at the points where the plotted contours enter the shaded region. It is understood that the boundary indicated by the top of the floor does not represent an abrupt change in the nature of the AGC response to jamming. The v_c transients still exist, but the jamming cycle frequency is so high that they have become mere ripples riding on the dc component of the control voltage. It is then that the dc component governs gain and, therefore, signal suppression.

Figure 6.9 indicates that the 20-dB J/S is more than adequate to disrupt the tracking function of the victim radar. Jamming with a 12% duty factor achieves 14 dB of signal suppression. The J/S based on average jamming power would then be

$$(J/S)_{av} = J/S + 10\log(\mathrm{DF}) = 20 - 9 = 11 \text{ dB}$$

Figures 6.10 and 6.11 were plotted to determine what values of signal suppression we might achieve at lower J/S. Figure 6.10 indicates that for DF = 12% we could achieve better than 9 dB of suppression with J/S = 15 dB. Figure 6.11 indicates that we could achieve a little over 4 dB of signal suppression with J/S = 10 dB.

The degree of signal suppression required to disrupt tracking depends on the tracking servo characteristics. A servo can be conditionally stable [4, Section 12.3], meaning that it becomes unstable if its loop gain drops appreciably below the design value. Even if instability does not result, the dynamic lag errors will increase as the gain is reduced, resulting in degraded tracking accuracy and possibly loss of tracking. Probably, one should aim for a value of signal suppression (and tracking servo loop gain reduction) of about 10 dB. Based on our linear AGC model, this would demand $J/S \approx 15$ dB.

We must understand that the results in Figures 6.9 through 6.11 are valid for a specific AGC model and for a specific signal level, $v_s = 0.001$ V. The plots change if v_s changes, even if J/S is held constant. Note that the higher values of signal suppression are achieved at low values of normalized period, P/T, and low values of DF. For a conical scan radar, the scan frequency might, for instance, be 30 Hz ($\omega_s = 2\pi f_s = 2\pi \times 30 = 188$ rad/s). At the end of Section 6.3 we indicated that the time constant T of the AGC loop filter should satisfy the following equation:

$$T > \frac{2\pi(1 + \mu\bar{v}_1)}{\omega_s}$$

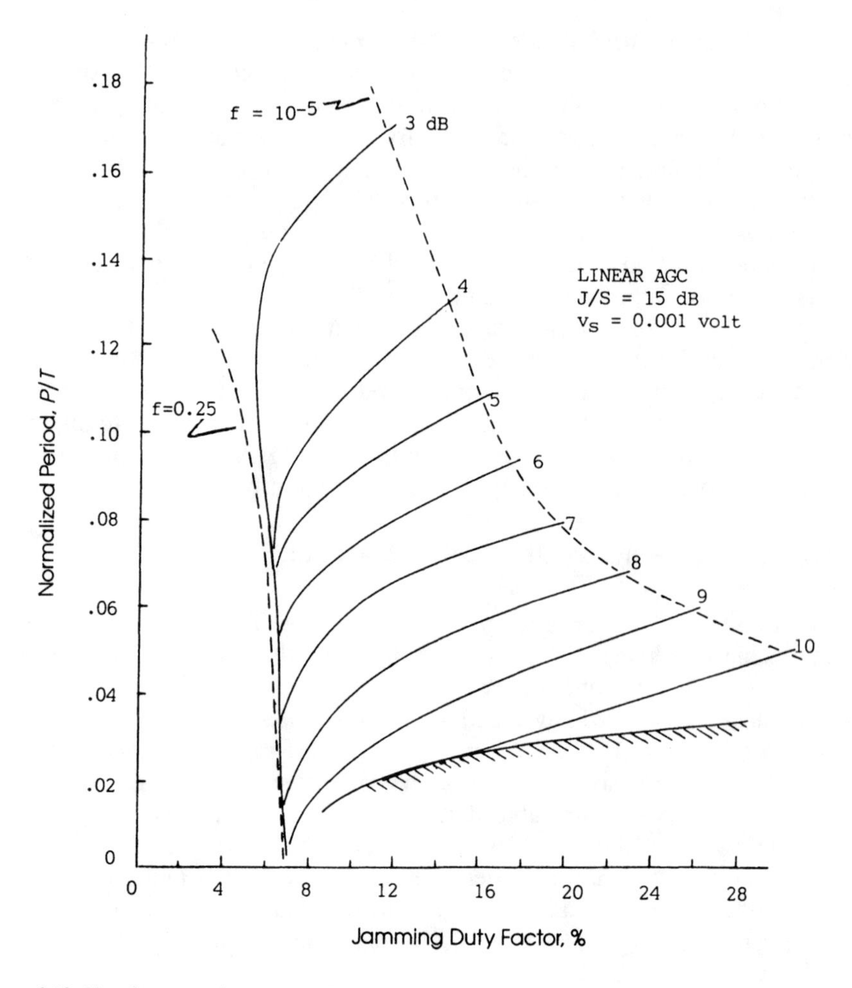

Figure 6.10 Signal suppression by on-off jamming.

With our illustrative parameters:

$$T > \frac{2\pi(1 + 10^5 \times 10^{-3})}{188} = 3.37 \text{ s}$$

Let us suppose that $T = 5$ s. If the jammer operates with $P/T = 0.03$, the jamming cycle period is $P = 0.15$ s. If the value of DF is 0.12, the jammer burst duration is only

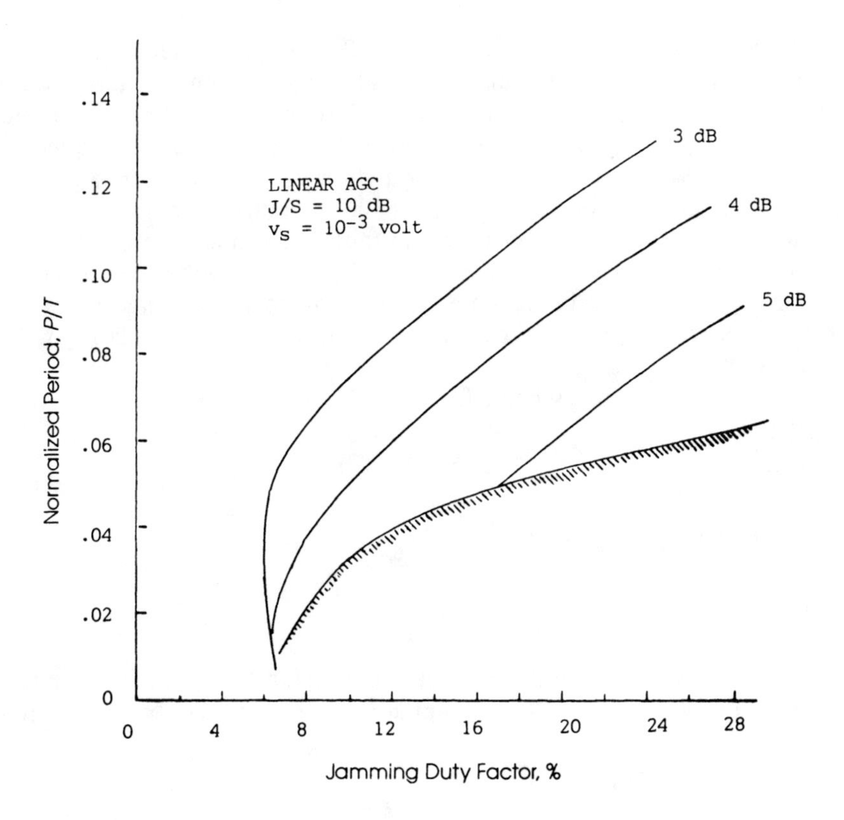

Figure 6.11 Signal suppression by on-off jamming.

$$T_{on} = P \times DF = 0.018 \text{ s}$$

The number of jamming pulses in the burst would be the integer nearest to

$$n_B = f_R T_{on}$$

where f_R is the PRF of the victim radar. If, for instance, the PRF = 1000 Hz:

$$n_B = 18 \text{ pulses}$$

and a jamming cycle encompasses a total of

$$m = f_R P = 150 \text{ pulses}$$

As a countermeasure against AGC deception, Barton [3, Chapter 10] suggests using fast AGC, with the target modulation signal (e.g., the conical scan modulation) extracted from the AGC voltage. By fast AGC we imply that the cut-off frequency of the AGC loop low-pass filter is well above the scan modulation frequency. The control voltage will then follow the modulation in the process of regulating the output of the gain-controlled amplifier to hold the output voltage essentially constant at v_r. Because of the short time constant of the filter, there are no long v_c transients of the sort that occur with slow AGC. Control voltage v_c rises to a high value during the jamming burst, but it drops rapidly to a mean level commensurate with the dc level of v_s and proceeds to reproduce the scan modulation when the jammer is off.

We can illustrate this scheme for an AGC with the logarithmic control law:

$$A = A_0 \beta_c^{-v_c}$$

by assuming an input signal of the form

$$v_1 = v_s(1 + \delta)$$

where $\delta = \delta(t) \ll 1.0$ is the scan modulation waveform. We shall assume that fast AGC holds output v_2 essentially constant at v_r, so the amplifier obeys the equation:

$$v_1 A = v_s(1 + \delta) A_0 \beta_c^{-v_c} = v_r$$

Setting $v_r = 1$ and taking the logarithms of both sides of the equation yields

$$\log v_s + \log(1 + \delta) + \log A_0 - v_c \log \beta_c = 0$$

For $\delta \ll 1.0$, $\log(1 + \delta) \approx \delta/(\ln 10)$, so the expression for v_c becomes

$$v_c = \frac{\log A_0 + \log v_s + \delta/(\ln 10)}{\log \beta_c}$$

The $\log v_s$ term is a dc term that adjusts the amplifier gain in accordance with input signal level. The $\delta/(\ln 10)$ term is the modulation component (e.g., $\delta = m \cos \omega_s t$) that can be extracted for use by the angle tracker. It is independent of signal level v_s, as is required if the tracking loop gain is to remain constant. One can easily show that an AGC obeying the linear control law would not be acceptable, for the modulation component of v_c is inversely proportional to v_s.

6.5 SIMULATION RESULTS

Simulations were devised as a check on the foregoing analysis of the linear model. The linear AGC was simulated by using the parameters of the preceding section:

$$A_0 = 10^4;$$
$$\mu = 10^5 \text{ V}^{-1};$$
$$v_r = 1.0 \text{ V};$$
$$v_{\text{lim}} = 2.5 \text{ V};$$
$$A_{\text{min}} = 5.0.$$

The simulated input signal was

$$v_1(t) = 0.001[1 + 0.1 \cos(120t)]$$

For slow AGC simulations, the filter time constant was set at $T = RC = 8$ s. For fast AGC simulations the value was $T = 0.05$ s.

Figure 6.12 is a plot of the linear slow AGC simulation for intermittent jamming with $J/S = 20$ dB, $P/T = 0.29$, and DF $= 0.12$. Figure 6.9 indicates a signal-suppression factor of about 12.5 dB for these parameters. This is the value we obtain by graphical integration of v_2 in Figure 6.12. The phase III transient of v_c is so short relative to the filter time constant that it appears linear.

Figure 6.13 is a simulation plot for the same parameters as in Figure 6.12, except that the normalized jamming cycle period is increased to $P/T = 0.155$ and DF is decreased to 0.099. This operating point on Figure 6.9 is well above the 6-dB contour, so all four v_c transient phases exist. In fact, we chose the point outside the enclosed region to produce a rather long phase IV component. The signal-suppression factor indicated by Figure 6.9 would be about 3.5 dB, and this is the value we obtain by graphical integration of Figure 6.13. The very rapid phase I transient is so short that it is difficult to see in Figure 6.13. It is clearly visible on the expanded plot of Figure 6.14, along with the phase II transient and the early part of phase III. This simulation demonstrates the importance of choosing jamming parameters that cut short or eliminate the phase IV transient in order to achieve a high signal-suppression factor.

Figure 6.15 is a plot of the fast AGC simulation. We can see only a faint trace of scan modulation in v_2. The dc and ac components of the v_c agree with the computed values. Note that v_c contains a large scan modulation component.

The logarithmic slow AGC was simulated with the following parameters:

$$A_0 = 10^4;$$
$$\beta_c = 10^{8.25};$$
$$v_r = 1.0 \text{ V};$$

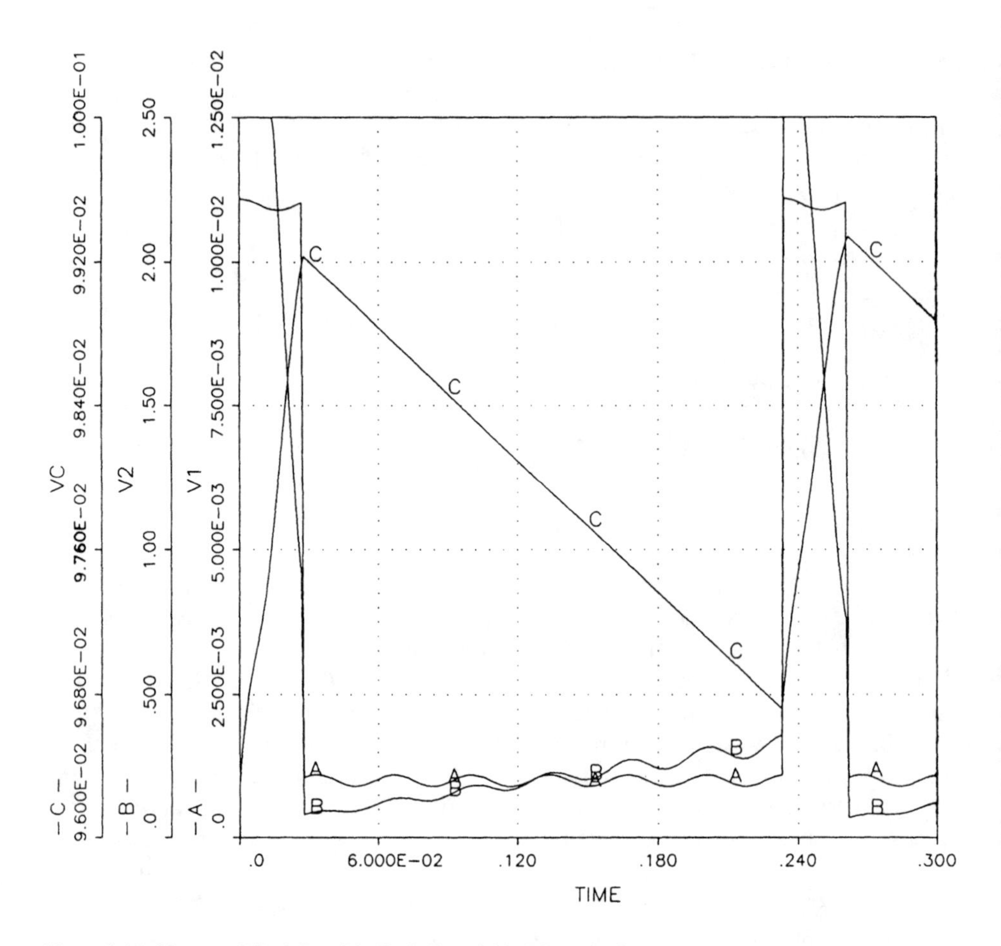

Figure 6.12 Linear AGC: J/S = 20 dB; P/T = 0.29; DF = 0.12.

$$v_{\text{lim}} = 2.5\ \text{V};$$
$$v_{c\max} = 0.40012\ \text{V (determines } A_{\min}\text{)};$$
$$T = RC = 8.0\ \text{s}.$$

The input signal was the same as for the linear AGC simulations. The J/S was 20 dB. Figure 6.16 is a plot of the logarithmic AGC response to jamming, with P/T = 0.057 and DF = 0.12. The signal-suppression factor, estimated from the plot, is about 11 dB. The linear AGC model for the same jamming parameters (see Figure 6.9) yields 10-dB suppression. Figure 6.17 is a plot of the response of the same AGC for P/T = 0.029 and DF = 0.12. The signal suppression measured from this plot is about 10.9 dB. The linear AGC model for the same jamming parameters (see Figure 6.9) yields 12.8-dB suppression. These differences in responses of the two models emphasize the importance of having a valid model of the radar when one assesses the effect of jamming.

6.6 SCAN-FREQUENCY COMPONENTS OF JAMMING BURST PATTERN

As all the simulation plots show, output voltage v_2 is at a high level during at least part of the jammer burst duration. We may approximate this portion of v_2 as a

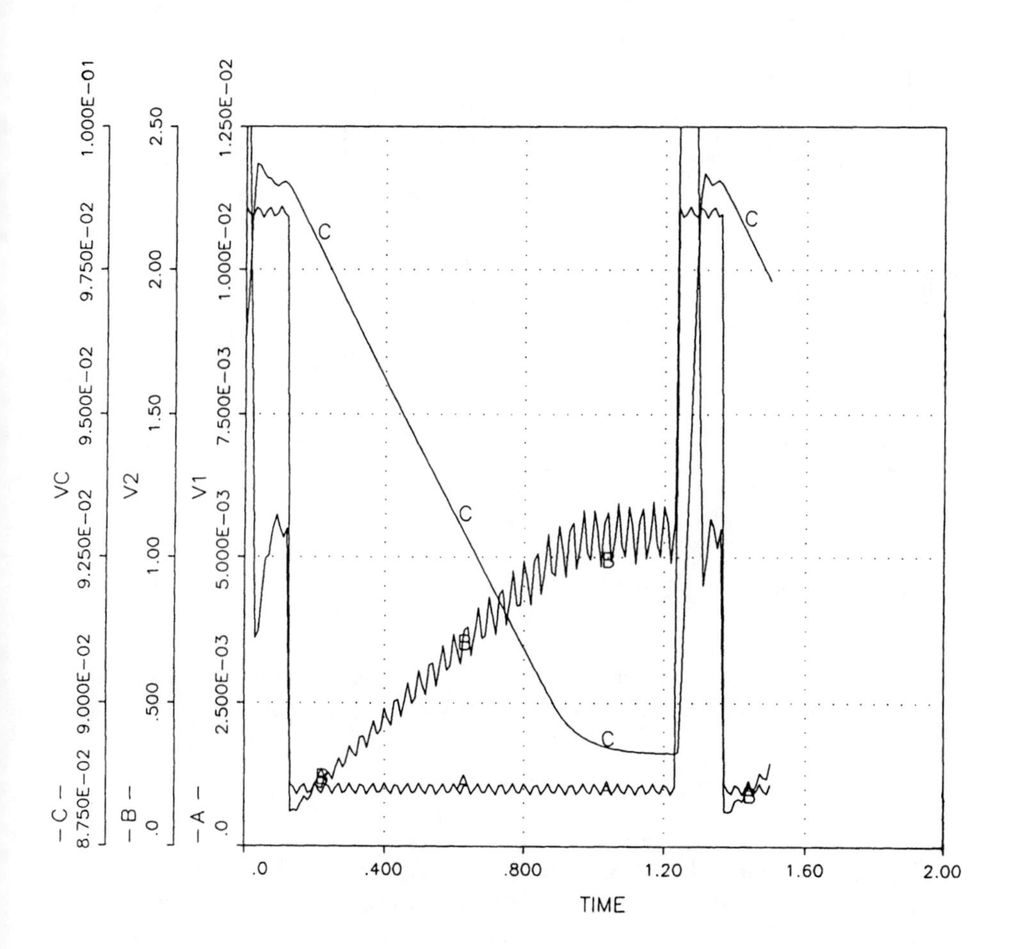

Figure 6.13 Linear AGC; $J/S = 20$ dB; $P/T = 0.155$; DF $= 0.099$.

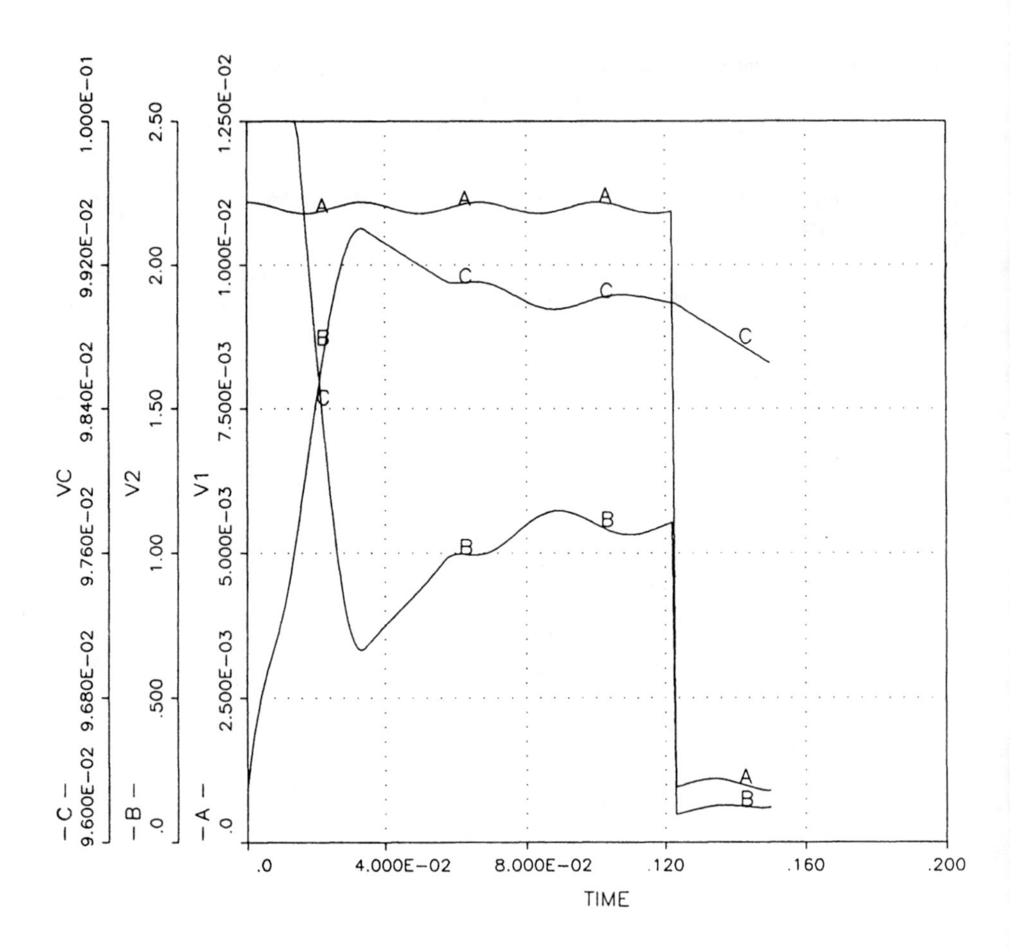

Figure 6.14 Expanded view of Figure 6.13.

train of rectangular pulses of period P and duration $P \times$ DF. If the pulse amplitude is v_{2j}, the amplitude of the nth harmonic component of the pulse train is

$$A_n = \frac{2v_{2j}}{\pi n} \sin(n\pi \mathrm{DF})$$

Possibly, a harmonic will fall near the CONSCAN frequency and appear to the tracker as a false target. From the point of view of the jammer, this would represent a bonus incidental to the signal-suppression effect.

The simulation of Figure 6.17 illustrates this phenomenon. The output voltage is at limit level, v_{lim}, during the entire burst, so $v_{2j} = v_{\mathrm{lim}} = 2.5$ V. The jamming parameters are P = 0.232 s, DF = 0.12, and f_s = scan frequency = 30 Hz. Clearly the seventh harmonic is very close to f_s:

$$\frac{7}{P} = 30.17 \text{ Hz}$$

Its amplitude is

$$A_7 = \frac{2 \times 2.5}{7\pi} \sin(0.84\pi) = 0.11 \text{ V}$$

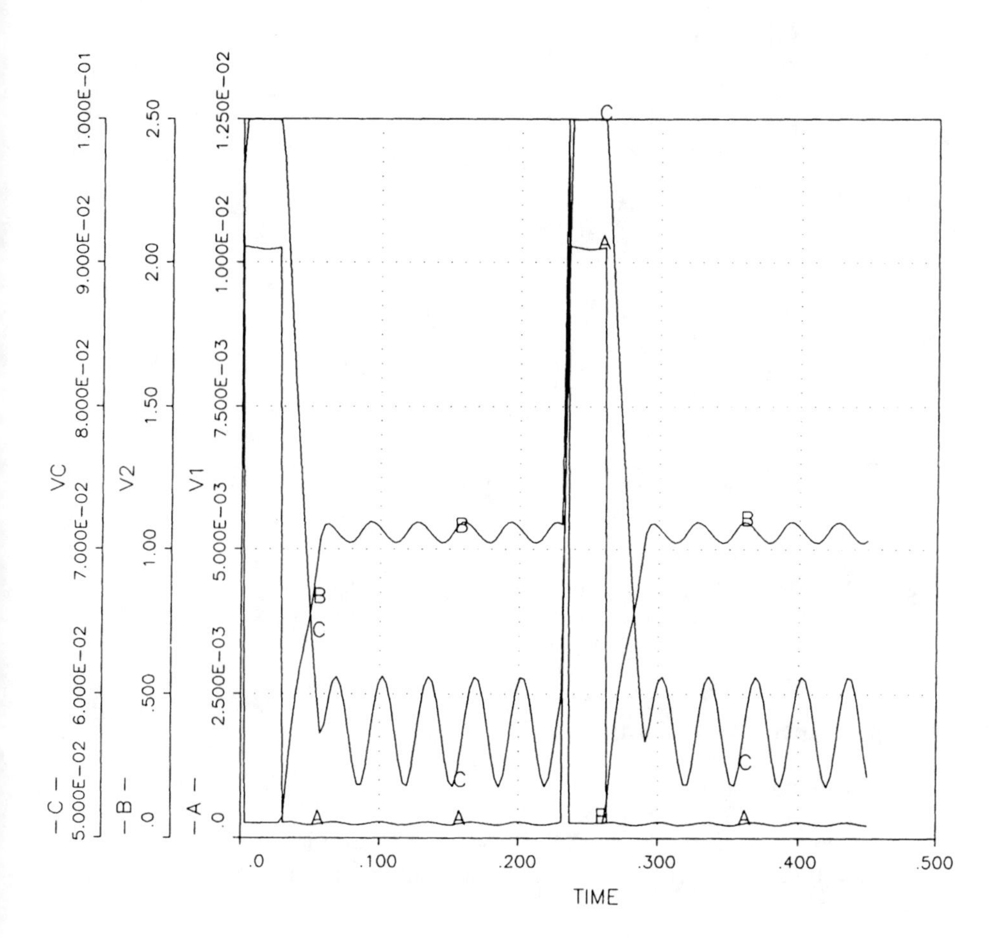

Figure 6.15 Linear fast AGC.

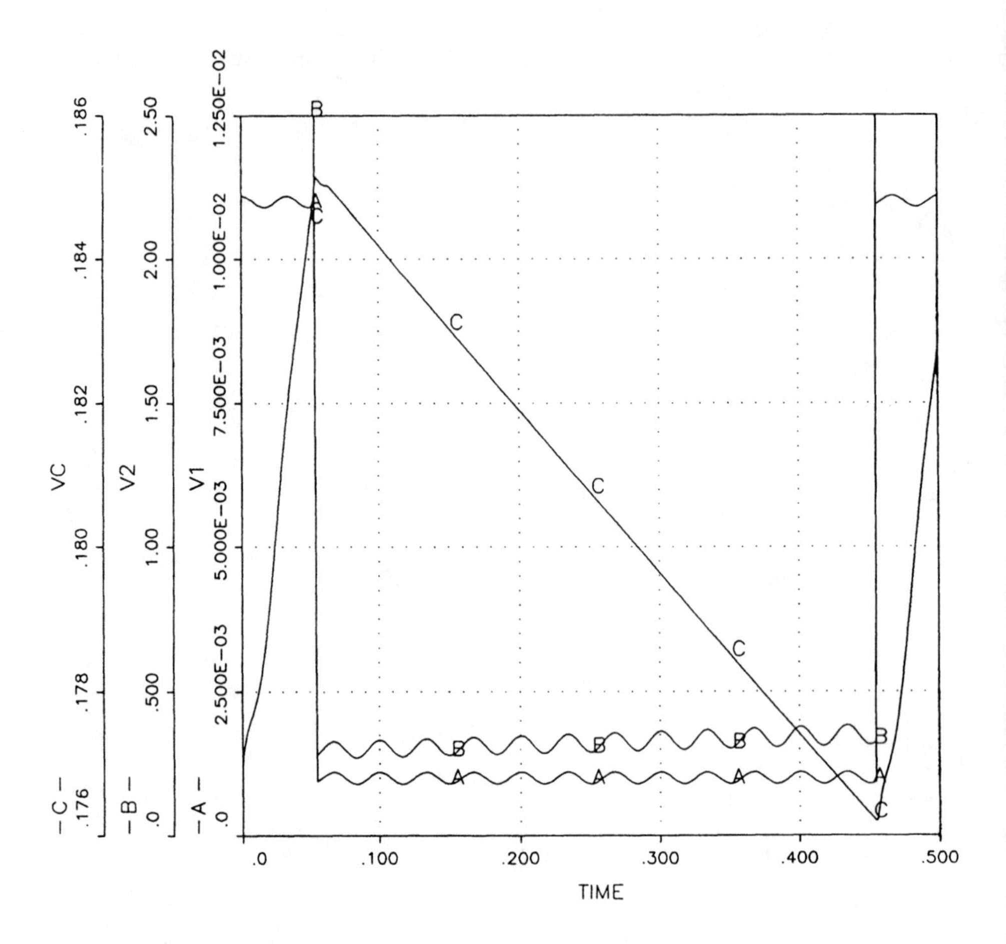

Figure 6.16 Logarithmic AGC: J/S = 20 dB; P/T = 0.057; DF = 0.12.

The output signal is of the form:

$$v_s[1 + m \cos(2\pi f_s t)]$$

v_s has been suppressed to a level 10.9 dB below reference level v_r, where v_r = 1.0 V. The modulation index is m = 0.1, so the amplitude of the signal modulation component is

$$v_{sm} = 0.0285 \text{ V}$$

Thus, the incidental jamming component A_7 is almost 12 dB above v_{sm}.

This incidental jamming is of the form discussed in Chapter 4. Because the seventh harmonic is not precisely equal to the value of f_s, the resultant false target will appear to move on a circle about the CONSCAN tracker's boresight (scan) axis at the difference frequency, which is only 0.15 rotation per second. It would appear likely that a jammer programmer, knowing the f_s of the victim radar, would choose a value of P that would place a harmonic of $1/P$ close enough to f_s to create tracking

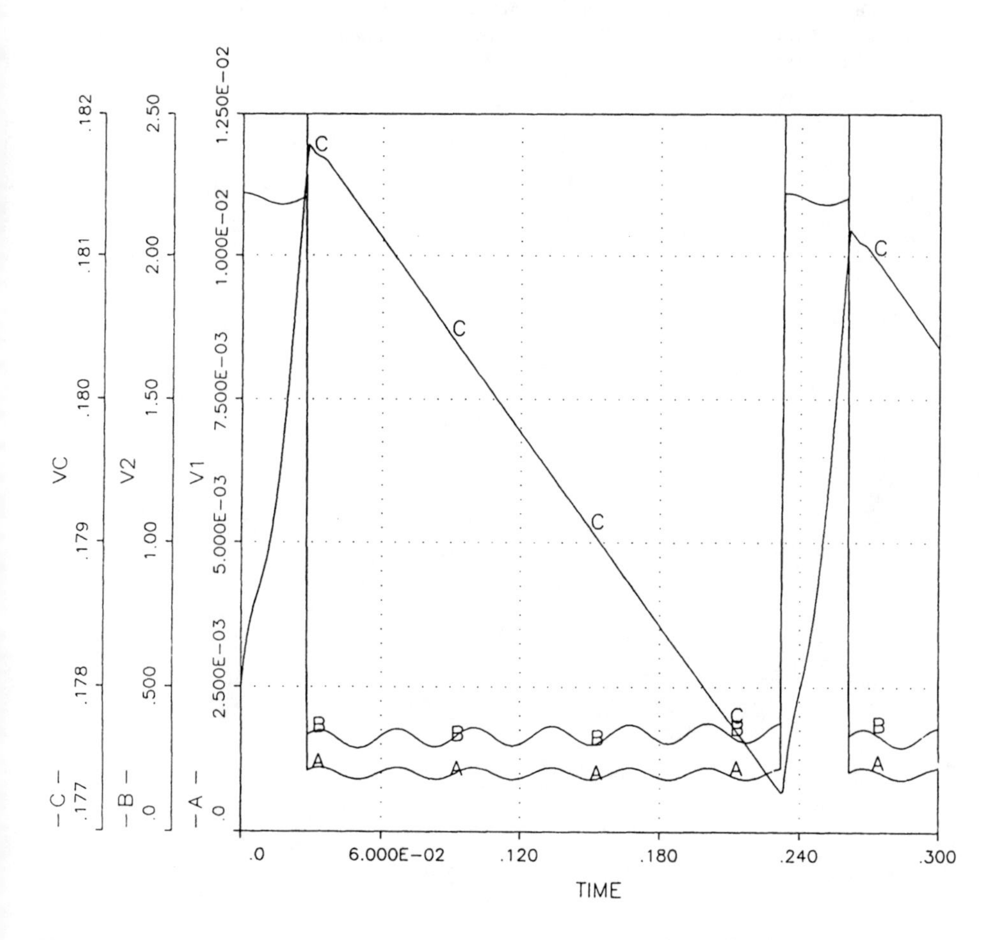

Figure 6.17 Jammer pattern's seventh harmonic matched to CONSCAN frequency.

disruption,[4] provided that the chosen P is compatible with the goal of AGC disruption (signal suppression). Figures 6.9 through 6.11 indicate that there can be considerable leeway in choosing P. We can manipulate the combination of P and DF to remain on a chosen signal-suppression contour.

REFERENCES

1. Hughes, R.S. *Analog Automatic Control Loops in Radar and EW,* Artech House, Norwood, MA, 1986.
2. Maksimov, M.V., *et al., Radar Anti-Jamming Techniques,* Artech House, Norwood, MA, 1979.
3. Barton, D.K., *Modern Radar System Analysis,* Artech House, Norwood, MA, 1988.
4. Chestnut, H., and R.W. Mayer, *Servomechanisms and Regulating System Design,* John Wiley, New York, 1951.

[4]The difference frequency needs to be low enough to cause false-target motion that the tracker can follow.

Appendix A
FEEDBACK CONTROL SYSTEMS

Because radar-based fire control systems rely on accurate tracking of the target, the tracking loops are likely to be subjected to jamming. Some knowledge of automatic control systems is essential for understanding the effects of jamming on the tracking loops. This appendix briefly reviews the principles of automatic control through feedback.

Background

The earliest closed-loop controllers (e.g., engine speed regulators (governors) and temperature regulators) were designed not on the basis of any theoretical foundations but through the application of common sense. The development of an elaborate feedback control theory grew out of the need for feedback amplifiers in telephone repeaters before World War II, but this theory did not appear in book form [1] until the end of the war. The war greatly stimulated the further development of control theory and extended its application into many new areas. Many books on the theory and application of automatic control became available shortly after the end of the war [2–4].

Tools for Analysis and Design

The control theory of most of these books presumes continuous linear systems, *continuous* meaning that the system variables are continuous functions of time, and *linear* meaning that system behavior is described by linear differential equations. Transform methods are used to systematize the solution of linear differential equations and to eliminate much tedious labor. The Laplace transform establishes a unique relation between a time function $f(t)$ and its transform $F(s)$. The transform variable, $s = \sigma + j\omega$, can be viewed as a complex frequency; e^{st} can represent rising or decaying exponentials if S is pure real; $e^{st} + e^{s^*t}$ can represent rising or decaying sinusoids (s^* is the complex conjugate of s). If the value of s is pure imaginary ($\sigma = 0$), the sinusoid has constant amplitude. A system's time response to an applied forcing function or to stored energy (initial conditions) is found as the inverse transform of the pertinent $F(s)$. The transform method owes its utility to the existence of extensive tables [5] of Laplace transform pairs. If a driving function $f_d(t)$ is applied to a system with a transfer function $G(s)$, the transform of the response is

$$F(s) = F_d(s)G(s)$$

where $F_d(s)$ is the transform of $f_d(t)$. For a lumped-constant system (as distinguished from a distributed parameter system), $G(s)$ is a rational function (ratio of two polynomials in s). The poles of $G(s)$ (i.e., the roots of the denominator polynomial) determine the characteristic response modes of the system. We obtain the steady-state frequency response by setting $s = j\omega$ in $G(s)$.

In a feedback system with nominally negative feedback, the feedback may become positive due to phase lags at higher frequencies. If the loop gain at these frequencies is greater than unity (0 db), the system becomes unstable. Various techniques have been developed for detecting potential instability.

The Nyquist stability theorem employs the behavior of $G(j\omega)$, as ω goes from $-\infty$ to $+\infty$, to detect the presence of poles of $G(S)$ in the right half of the s-plane (such poles correspond to unbounded exponentially rising time functions). The theorems of Bode [1] also employ $G(j\omega)$ to characterize the behavior of feedback control systems. Plots of $\log|G(j\omega)|$ *versus* $\log \omega$ and $\arg[G(j\omega)]$ as a function of ω are called *Bode plots.* They are extremely useful tools for design, particularly because we can draw the asymptotes of $\log|G(j\omega)|$ by inspecting $G(s)$ in its factored form (the factored form of $G(s)$ reveals the pole and zero locations). Likewise, quick approximations for $\arg[G(j\omega)]$ are available. These plots enable quick determination of phase and gain margins of a given design. Phase margin is the amount by which the phase, $\arg[G(j\omega)]$, falls short of 180° when ω reaches the value producing a gain of 1.0, or 0 dB, in the magnitude $|G(j\omega)|$. The gain margin is the amount by which $|G(j\omega)|$ falls below 0 dB at the frequency at which $\arg[G(j\omega)]$ has reached $-180°$. The use of Bode plots is covered in standard texts on control theory [2].

As Laplace transforms facilitate the study of continuous linear systems, the z-transform facilitates the study of linear sampled-data systems. Pulsed radar obtains discrete samples of the target variables. Often we can convert these sample sequences, by sample-and-hold operations and by filtering, into continuous replicas of the original continuous-time functions, in which case the control system operates on continuous inputs. There are systems in which processing of the input signals is based entirely on discrete samples (digital processing), and even the control of an output variable may occur as discrete steps (stepping motors).

Most real systems contain some nonlinearities, but if the nonlinear effects are small a linear model provides an adequate representation of system behavior. Jamming, especially against tracking loops, aims to produce large tracking errors, causing the error detectors to become decidedly nonlinear. Linear analysis then fails, and one is forced to simulation, in which the nonlinearities can be accounted for. For about three decades after World War II, electrical analog computers were used for simulating linear and nonlinear systems. Beginning in the third decade, digital simulation came into being, and for a time hybrid analog-digital simulators were in use. In recent years essentially all simulation has been performed digitally. The integration of differential equations is performed numerically, with the integration step size made small enough to ensure that this discrete process yields an accurate representation of the corresponding continuous process.

Transforms and Transfer Functions

The transfer function $G(s)$ is the generalized counterpart of the concepts of impedance, admittance, and the steady-state frequency transfer function $G(j\omega)$. These latter concepts arise out of the solutions of the system differential equations in which the forcing function is a continuously applied sinusoid of angular frequency ω. The fundamental approach for deriving a transfer function is to go from the system differential equations to the transform equations by taking Laplace transforms term by term. Initial conditions, representing stored energy, can be included in the resultant transform equation. Table A.1 is a short list of transform pairs. The time function is $f(t)$, and its Laplace transform is

$$L[f(t)] = F(s)$$

Conversely, $f(t)$ is the inverse transform of $F(s)$:

$$L^{-1}[F(s)] = f(t)$$

The argument $(0+)$ appearing in Table A.1 represents the instant at the beginning of the solution, $f'(0+)$ represents the initial value of the first derivative of $f(t)$, $f^{(-1)}(0+)$ represents the initial value of $\int f(t)\,dt$, and $f^{(-2)}(0+)$ represents the initial value of $\int\int f(t)\,dt^2$.

With relaxed initial conditions ($v_c(0+) = 0$) the transform of the differential equation:

$$v_1(t) = Ri(t) + \frac{1}{C}\int_0^t i\,dt$$

for the circuit of Figure A.1 becomes

Table A.1
Some Laplace Transform Pairs

$f(t)$	$F(s)$	
$af(t)$	$aF(s)$	linearity
$f_1(t) + f_2(t)$	$F_1(s) + F_2(s)$	linearity
$\dfrac{df(t)}{dt}$	$sF(s) - f(0+)$	differentiation
$\dfrac{d^2f(t)}{dt^2}$	$s^2F(S) - sf(0+) - f'(0+)$	differentiation
$\int_0^t f(t)\, dt$	$\dfrac{F(s)}{S} + \dfrac{f^{(-1)}(0+)}{S}$	integration
$\int\int_0^t f(t)\, dt^2$	$\dfrac{F(s)}{S^2} + \dfrac{f^{(-1)}(0+)}{S^2} + \dfrac{f^{(-2)}(0+)}{S}$	integration
$\delta(t)$	1	unit impulse at $t = 0$
$u(t)$	$\dfrac{1}{s}$	unit step at $t = 0$

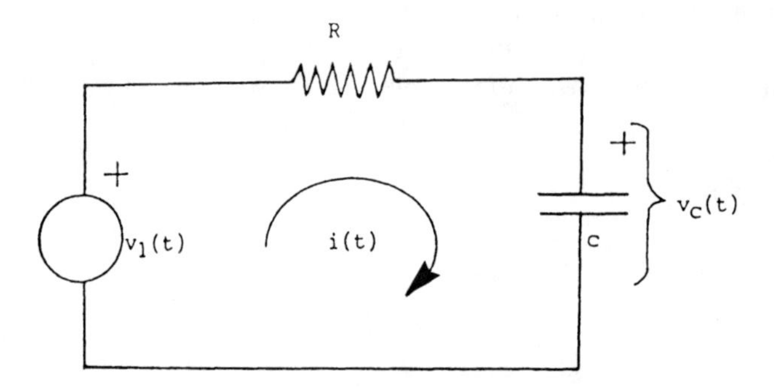

Figure A.1 Simple series circuit.

$$V_1(s) = I(s)\left(R + \frac{1}{Cs}\right)$$

The transform of the current is

$$I(s) = V_1(s)\,\frac{1}{R + 1/Cs}$$

If $v_1(t)$ is a step voltage of magnitude V, the transform of the driving function is V/s, and reference to transform tables will yield the familiar current waveform:

$$i(t) = \frac{V}{R} e^{-t/RC} \qquad t \geq 0$$

The factor $1/(R + 1/Cs)$, which is the current-voltage transfer function, reduces to the familiar steady-state circuit admittance when s is replaced by $j\omega$:

$$Y(j\omega) = \frac{I(j\omega)}{V_1(j\omega)} = \frac{1}{R + 1/j\omega C}$$

If one is interested in the capacitor voltage rather than the current, the appropriate transfer function is

$$G(s) = \frac{V_C(s)}{V_1(s)} = \frac{1}{sRC + 1}$$

This is the simple, single-pole filter transfer function (pole at $s = -1/RC$), which reduces to the steady-state frequency transfer function:

$$G(j\omega) = \frac{1}{j\omega RC + 1}$$

for $s = j\omega$.

Feedback control systems consist of numerous interconnected parts. Consequently, we need higher-order differential equations or sets of low-order equations to describe their behavior, resulting in transfer functions with higher-degree polynomials in s in their denominators and numerators.

Terminology

Figure A.2 depicts a feedback control system in which the output variable $y(t)$ is forced to follow the input variable $x(t)$ by application of feedback of the proper sense to drive the error

$$e(t) = x(t) - y(t)$$

toward zero. The open-loop transfer function is

$$G(s) = \frac{Y(s)}{E(s)} = \frac{Y(s)}{X(s) - Y(s)}$$

and the closed-loop transfer function is

$$H(s) = \frac{Y(s)}{X(s)} = \frac{G(s)}{1 + G(s)}$$

Such a system is generally called a *regulator* if its purpose is to cause the output y to remain close to a fixed value of x in spite of external disturbances. Regulator requirements therefore differ from those of the systems in which y must follow a dynamically changing input. A feedback system in which there is dynamic control of a mechanical output position is called a *servomechanism.*[1] Because mechanical and electromechanical components may be less familiar to the reader than are electric circuit components. We briefly review mechanical principles and terminology.

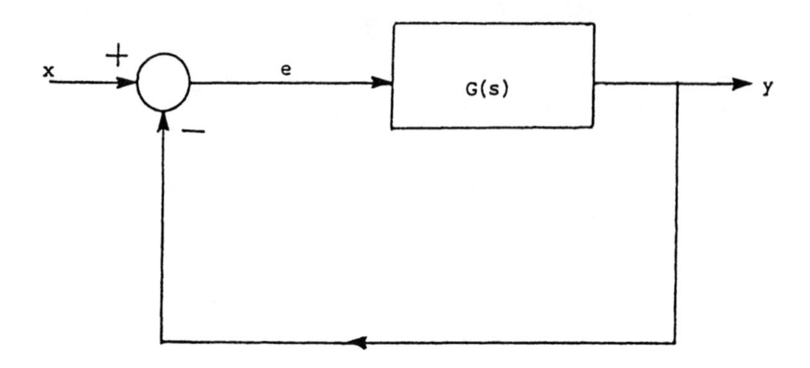

Figure A.2 Simple feedback control loop.

Mechanical Components

Linear electric networks and linear mechanical systems are described by equations of the same form. For electric networks, the variables are voltages, currents, and their derivatives and integrals. The coefficients involve resistances, inductances, and capacitances. In translational mechanical systems, the variables are forces, displacements, and their derivatives and integrals. The coefficients involve spring stiff-

[1]This terminology is not rigidly followed. The term *servomechanism* or *servo* is often applied to other kinds of feedback control systems.

ness constants, viscous friction constants, and masses. In rotational mechanical systems, the variables are torques, rotational displacements, and their derivatives and integrals. The coefficients involve torsion spring stiffness constants, rotational friction constants, and moments of inertia. Figure A.3 gives the basic force and motion relations for mechanical components. In the columns containing dimensions, the brackets are used to signify dimensions of the bracketed entity.

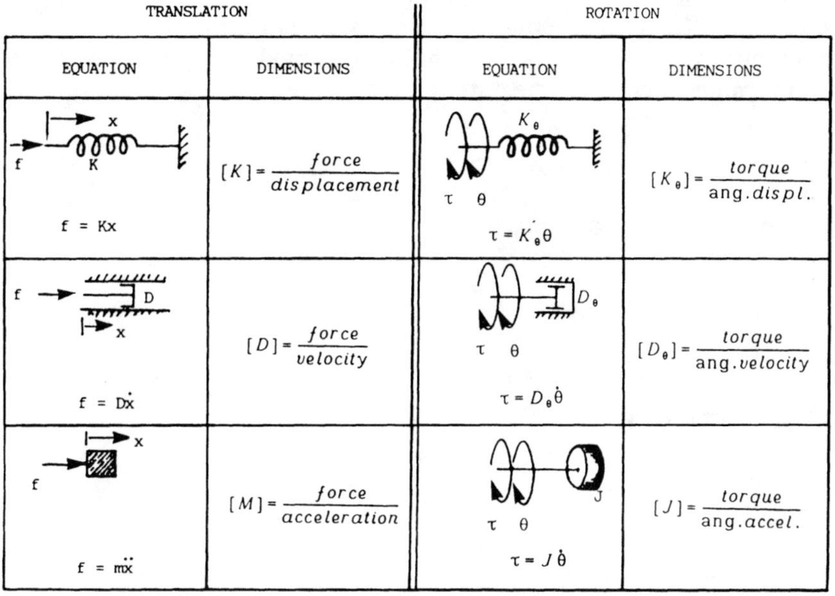

TRANSLATION		ROTATION	
EQUATION	DIMENSIONS	EQUATION	DIMENSIONS
x, f, K $f = Kx$	$[K] = \dfrac{force}{displacement}$	K_θ, τ, θ $\tau = K_\theta \theta$	$[K_\theta] = \dfrac{torque}{ang.displ.}$
f, D, x $f = D\dot{x}$	$[D] = \dfrac{force}{velocity}$	D_θ, τ, θ $\tau = D_\theta \dot{\theta}$	$[D_\theta] = \dfrac{torque}{ang.velocity}$
x, f $f = m\ddot{x}$	$[M] = \dfrac{force}{acceleration}$	J, τ, θ $\tau = J\ddot{\theta}$	$[J] = \dfrac{torque}{ang.accel.}$

Figure A.3 Components of mechanical systems.

In Figure A.3, force has been chosen as one of the fundamental entities, the other two being length and time. If the dimensions of these fundamental entities are indicated by F, L, and T, the dimensions of velocity and acceleration are, respectively, LT^{-1} and LT^{-2}, and mass has the dimensions $FL^{-1}T^2$. In the rotational system the angular displacement is dimensionless, and angular velocity and acceleration dimensions are, respectively, T^{-1} and T^{-2}. Torque has the dimensions FL, and moment of inertia has the dimensions FLT^2. Two common components, not included in Figure A.3, are levers (force-displacement transformers) and gear boxes (torque-displacement transformers). The equations for these components are easily derived in terms of the lever ratio or gear ratio.

Example: A Simple Position Controller: We begin by assuming a servomechanism with block diagram corresponding to Figure A.2. This servo, depicted in Figure A.4(a), controls the azimuth pointing angle, θ_A, of an antenna, causing the antenna to follow the target azimuth angle, θ_T. The error angle is $\epsilon = \theta_T - \theta_A$. An error transducer converts this to a voltage $K_\epsilon (\theta_T - \theta_A)$. A tachometer on the output shaft provides an output velocity-damping voltage, $K_D\dot{\theta}_4$ that is added to the error voltage with proper polarity at the input to the amplifier. The amplifier output voltage is applied to the armature of the antenna servo motor. The armature resistance lumped together with an amplifier output resistance is R_m. The back EMF constant[2] of the motor is K_B, and the motor torque constant is K_τ. Figure A.4(b) shows the linear speed-torque relationship of the motor. This model neglects armature inductance and ignores friction on the output shaft.

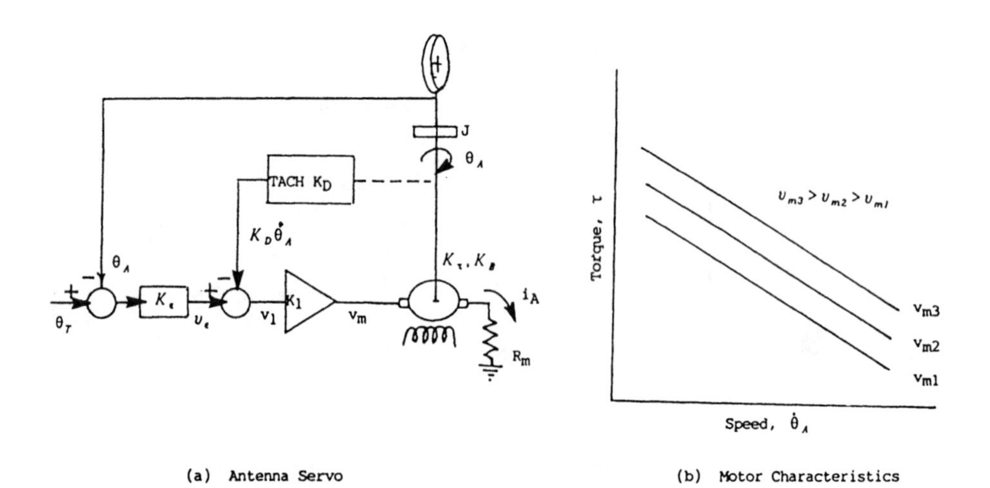

Figure A.4 Antenna servo description.

The following equations govern the operation of the controller:

$$\epsilon = \theta_T - \theta_A$$

$$v_\epsilon = K_\epsilon \epsilon$$

$$v_1 = v_\epsilon - K_D\dot{\theta}_A$$

[2]This constant relates induced *electromotive force* (EMF) to motor speed. Bruns and Saunders [8] describe the characteristics of various servomotors.

$$v_m = k_1 v_1$$

$$v_m = R i_A + K_B \dot{\theta}_A$$

$$\tau_m = K_\tau i_A$$

$$\tau_m = J \ddot{\theta}_A$$

With relaxed initial conditions, the transform equations are (uppercase symbols are used to represent the transforms of the corresponding time functions)

$$E(s) = \Theta_T(s) - \Theta_A(s)$$

$$V_\epsilon(s) = K_\epsilon E(s)$$

$$V_1(s) = V_\epsilon(s) - K_D s \Theta_A(s)$$

$$V_m(s) = K_1 V_1(s)$$

$$V_m(s) = R_m I_A(s) + K_B s \Theta_A(s)$$

$$\tau_m(s) = K_\tau I_A(s)$$

$$\tau_m(s) = J s^2 \Theta_A(s)$$

From these equations, we find the open-loop transfer function:

$$G(s) = \frac{\Theta_A(s)}{E(s)} = \frac{K_1 K_\epsilon K_\tau / R_m J}{s[s + (K_1 K_D + K_B) K_\tau / R_m J]}$$

The closed-loop transfer function is

$$H(s) = \frac{\Theta_A(s)}{\Theta_T(s)} = \frac{G(s)}{1 + G(s)}$$

$$H(s) = \frac{K_1 K_\epsilon K_\tau / R_m J}{s^2 + (K_\tau / R_m J)(K_1 K_D + K_B) s + K_1 K_\epsilon K_\tau / R_m J}$$

This transfer function is in the classical form:

$$H(s) = \frac{\omega_n^2}{s^2 + 2\zeta \omega_n s + \omega_n^2}$$

In this transfer function, ω_n is the undamped natural frequency and ζ is the damping ratio. Oscillation occurs at ω_n only if $\zeta = 0$. With finite damping, the frequency[3] of the damped oscillation is $\omega_n\sqrt{1 - \zeta^2}$.

The response to a unit step is obtained by setting $\Theta_T(s) = 1/s$, yielding

$$\Theta_A(s) = \frac{\omega_n^2}{s(s^2 + 2\zeta\omega_n s + \omega_n^2)}$$

The time response is (for $t \geq 0$)

$$\Theta_A(t) = 1 - \frac{e^{-\zeta\omega_n t}}{\sqrt{1 - \zeta^2}} \sin(\omega_n\sqrt{1 - \zeta^2}\, t - \psi)$$

$$\psi = \tan^{-1}\left[\frac{-\sqrt{1 - \zeta^2}}{\zeta}\right]$$

The following equations relate ω_n and ζ to the parameters of Figure A.4:

$$\omega_n^2 = \frac{K_1 K_\epsilon K_\tau}{R_m J}$$

$$\zeta = \frac{(K_1 K_D + K_B)K_\tau}{2\omega_n R_m J}$$

Let us proceed through a preliminary design in which the approximate physical parameters of the antenna and its mount are known. Figure A.5 represents the reflector as an ellipse of dimensions $W \times H \times t_a$, and the mounting base as a circular disk of diameter D and thickness t_b. The reflector is aluminum ($\sigma_a = 165$ lb/ft^3), and the base is steel ($\sigma_b = 480$ lb/ft^3). The moment of inertia of the combination about the rotation (vertical) axis is

$$J = \frac{1}{g} \times \frac{\pi}{64}\,[W^3 H t_a \sigma_a + D^4 t_b \sigma_b]$$

where g = acceleration of gravity (32.2 ft/s^2).

The dimensions in Figure A.5 correspond to an airborne radar antenna. With those dimensions, the moment of inertia is[4] $J = 0.344$ slug ft^2. The undamped

[3]As used here, the term *frequency* means angular frequency (radians per second).

[4]The English units of this example can be readily converted to MKS units. Refer to conversion tables in any handbook [5].

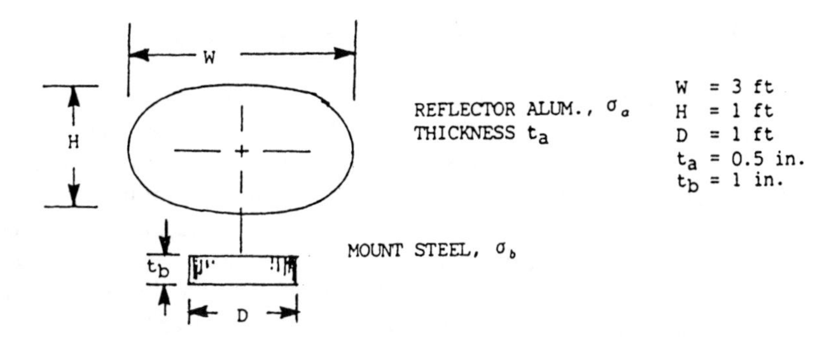

Figure A.5 Antenna model.

natural frequency is specified at $\omega_n = 10$ rad/s, and the damping ratio is to be $\zeta = 0.5$. This will result in 16% overshoot in the servo step response.[5] The selected motor has the following parameters: $K_\tau = 0.1$ ft·lb/A, $R_m = 0.1\ \Omega$, $K_B = 0.3$ V/(rad/s), and the error transducer constant is $K_\epsilon = 2$ V/rad. The requirement on ω_n then dictates the amplifier voltage gain, K_1:

$$K_1 = \frac{\omega^2 R_m J}{K_\tau K_\epsilon} = 17.2$$

Only the tachometer feedback constant, K_D, remains to be determined. We find it by using the requirement on the damping ratio:

$$K_D = \frac{1}{K_1}\left[\frac{2\zeta\omega_n R_m J}{K_\tau} - K_B\right] = 0.183 \text{ V/(rad/s)}$$

This system illustrates the advantage of electronic damping over mechanical damping. The former is simply set by adjusting the voltage gain in the velocity ($\dot{\theta}_A$) feedback path. This damping results in no power dissipation. The same damping ratio can be achieved by a rotational damper (e.g., rotating vanes in a viscous fluid bath) on the output shaft, but this would result in added weight and it would dissipate power whenever the output shaft is in motion.

With the preceding parameters, the open-loop transfer function is

$$G(s) = \frac{100}{s(s + 10)}$$

[5]For this value of ζ the transient error falls below 2.5% in a nominal settling time of $t_s = 7.5/\omega_n$.

The plot in Figure A.6 of the unit-step response of this system was obtained by simulation.

We should bear in mind that a number of details have been slighted in this hasty preliminary design. We assumed a direct-drive motor, whereas it is quite possible that the antenna would be driven through a gear box, but this is easily accounted for. The description of the system implied that the moment of inertia of the rotor of the drive motor is negligible. This is unlikely to be the case, but the rotor inertia would simply be lumped with the load inertia. We included no bearing or gear friction. Those friction components are not realistically representable as ideal viscous friction. Their effect is likely to be small compared with the damping torque produced by the velocity feedback damping, so they are neglected. We neglected the rotor inductance, L_m, of the motor. If we had included its effect, $G(s)$ would exhibit another pole. The effect is demonstrated by adding a small L_m in

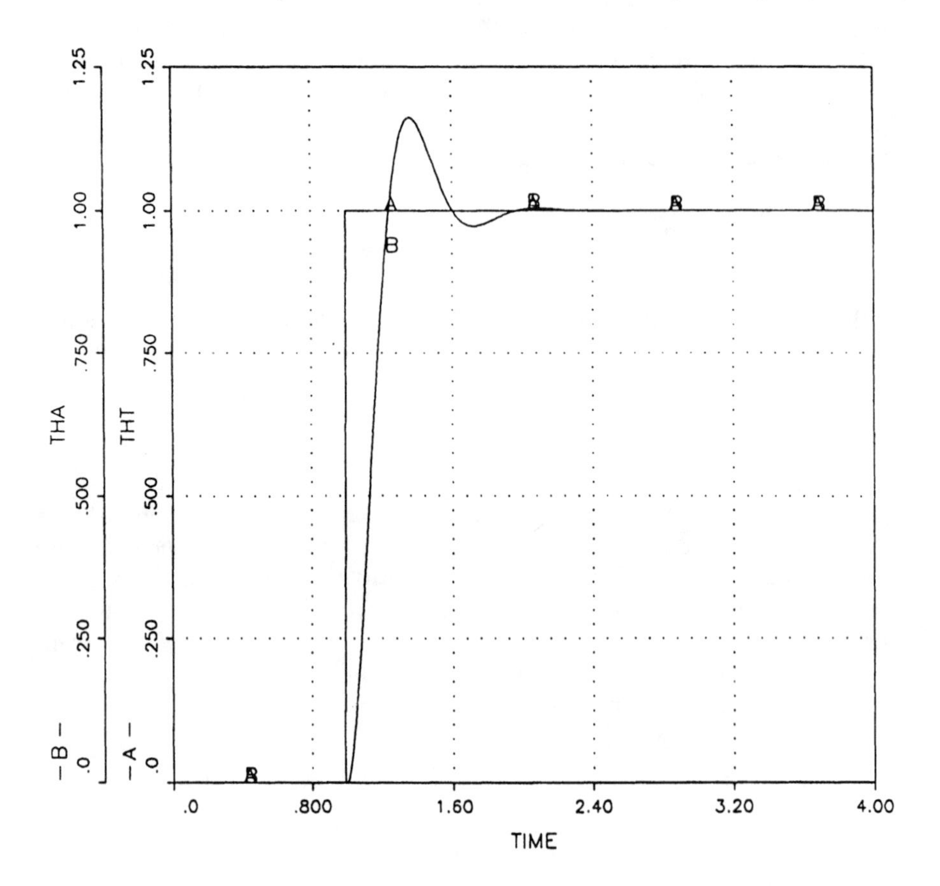

Figure A.6 Type 1 servo step response (ω_n = 10 rad/s; ζ = 0.5).

series with R_m. If we increase L_m from 0 to 1.0 mH $= 10^{-3}$ H, the existing pole is shifted from $s_1 = -10\ s^{-1}$ to $s_1 = -11.3\ s^{-1}$, and the new pole appears at $s_2 = -85\ s^{-1}$. That this new pole has negligible impact on system response can be demonstrated on a Bode plot [2–4] of $G(s)$.

Figure A.7 is a Bode plot for the system with L_m neglected. The breakpoint between the −20-dB/decade and −40-dB/decade slopes of $|G(j\omega)|$ is at $\omega = \omega_n = 10$ rad/s. The crossover (unity gain) frequency is at $\omega = 7.85$ rad/s. The phase lag at this frequency is 128°, so there is a positive phase margin of 180° − 128° = 52°, resulting in good stability (the damping ratio was set at $\zeta = 0.5$). As we noted, inserting the 1.0-mH inductance raises the frequency of the first breakpoint only slightly (from 10 to 11.3 rad/s). The second pole would result in a second breakpoint in the $|G(j\omega)|$ at $\omega = 85$ rad/s. The phase lag at gain crossover contributed by this second pole is only about 6°, so there is little change in stability or transient response. If the inductance were appreciably greater than a millihenry, its effect would have to be taken into account. In fact, as L_m approaches 2.5 mH, the two poles approach each other along the negative real axis, reaching coincidence (double pole) at $s = -20\ s^{-1}$. As L_m increases beyond 2.5 mH, the two poles diverge and become complex conjugates, indicating a resonance in $G(j\omega)$.

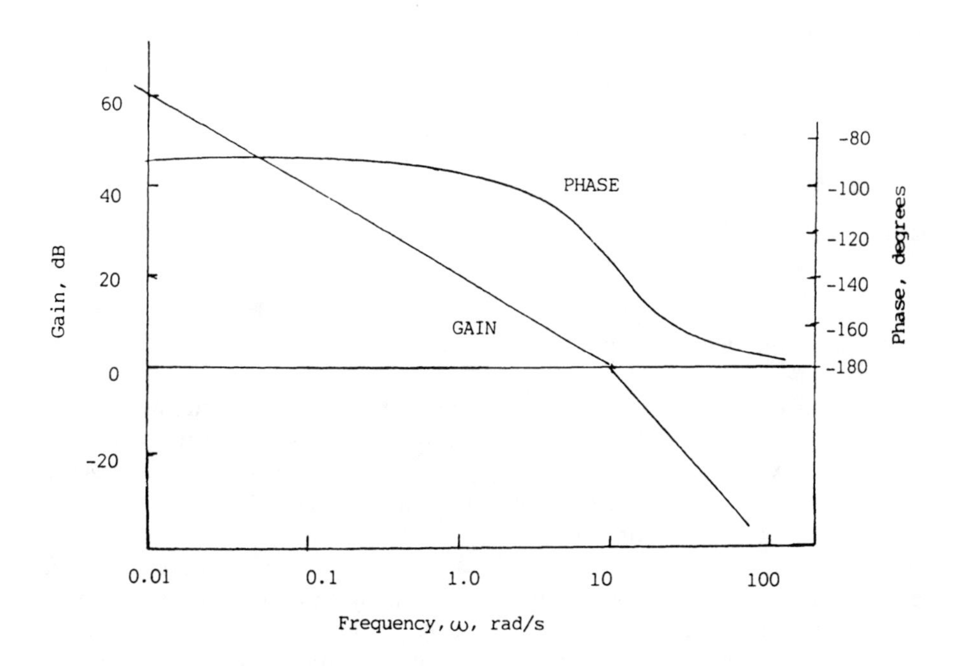

Figure A.7 Bode plot for antenna servo example.

One other approximation deserves mention. The error detector was modeled as a linear device with constant K_ϵ. In reality, when the measurement of pointing error is derived from the radar return itself, the error detector (e.g., monopulse or conical scan error detector) is linear only for small angular errors. As Figure A.8 indicates, the detector curve is typically in the form of an S curve. The linear approximation is adequate for studying normal tracking behavior (small angular errors), but in jamming situations the jammer's objective is to create large errors. A realistic model must then include the true detector characteristic.

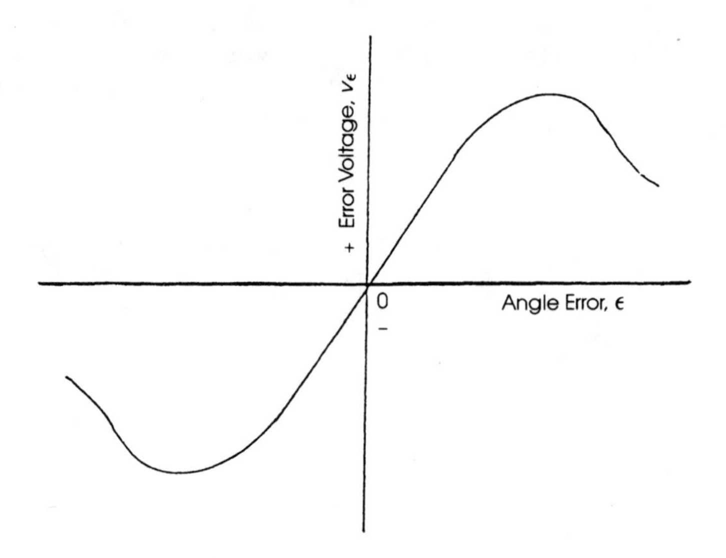

Figure A.8 Typical angle error detector characteristic.

Servo Classification and Static Error Coefficients

Position control systems are required to follow inputs that are sometimes constant and sometimes changing at constant rates or with constant accelerations. For instance, an azimuth-tracking servo experiences a constant input (the target azimuth is constant) when the target moves along a radial path. For a passing course, with the target moving at constant speed v, there is a peak in the angular rate equal to v/R_0 as the target passes the point of nearest approach (minimum range $= R_0$). Generally, the duration of the nearly constant angular rate is long compared with the servo response time, so we can evaluate the tracking error for this situation by assuming a constant-angular-velocity input. For this same passing course, the angular acceleration of the line of sight to the target peaks to a value of $0.65(v/R_0)^2$ when the azimuth is $\pm 30°$ relative to the azimuth of nearest approach. We can evaluate the servo error resulting from such input accelerations by assuming this value as a constant-acceleration input.

For the single-loop system of Figure A.2, the classical static error coefficients are defined [7] as follows:

$$K_p = \text{position error coefficient} = \lim_{s \to 0} G(s)$$
$$K_v = \text{velocity error coefficient} = \lim_{s \to 0} sG(s)$$
$$K_a = \text{acceleration error coefficient} = \lim_{s \to 0} s^2 G(s)$$

These error coefficients, when finite, are equal to the gain constant K of the open-loop transfer function $G(s)$, when $G(s)$ is written in the following form:

$$G(s) = \frac{KA(s)}{s^n B(s)}$$

$$A(s) = (s\tau_1 + 1)(s\tau_2 + 1)(s\tau_2 + 1) \cdots$$

$$B(s) = (s\tau_a + 1)(s\tau_b + 1)(s\tau_c + 1) \cdots$$

The exponent n determines the control system type:

$n =$ 0: type 0;
$n =$ 1: type 1;
$n =$ 2: type 2.

The exponent n is equal to the number of integrators in the loop. For $n > 2$, the phase lag contributed by the integrators (even at frequencies approaching zero) is 270° or more, and it becomes difficult to stabilize the system.

It is clear that the only finite nonzero error coefficients are:

$K_p =$ gain constant K for type 0 system;
$K_v =$ gain constant K for type 1 system;
$K_a =$ gain constant K for type 2 system.

These finite coefficients are useful for evaluating steady-state position errors in three specific cases. The following equation relates the transform of the error, $E(s)$, to the transform of the input, $X(s)$, for the system of Figure A.2:

$$E(s) = \frac{X(s)}{1 + G(s)}$$

The final value theorem [5] states that

$$\lim_{t \to \infty} f(t) = \lim_{s \to 0} sF(s)$$

where $F(s)$ is the Laplace transform of $f(t)$. Applying this theorem to find the steady-state position error, e_{ss}, yields

$$e_{ss} = \lim_{t \to \infty} e(t) = \lim_{s \to 0} \frac{sX(s)}{1 + G(s)}$$

For the type 0 system with a unit-step displacement input, $X(s) = 1/s$:

$$e_{ss} = \frac{1}{1 + k_p}$$

For the type 1 system with a unit ramp (unit step of velocity), $X(s) = 1/s^2$:

$$e_{ss} = \frac{1}{K_v}$$

For the type 2 system with a unit step of acceleration as the input, $X(s) = 1/s^3$:

$$e_{ss} = \frac{1}{K_a}$$

The steady-state error is zero for a step position input for system types 1 and 2. It is also zero for a velocity step into a type 2 system. The steady-state error becomes infinite for velocity or acceleration steps into a type 0 system. It is also infinite for an acceleration step into a type 1 system. Generalized error coefficients [2,7] have been defined that provide more information on servo performance than do the static error coefficients defined here, but their application is beyond the scope of this appendix.

Compensation Networks

In the antenna servo example we found that the open-loop transfer function $G(s)$ was constrained by hardware parameters, but two electronically adjustable parameters were available: (1) the amplifier gain K_1 was selected to provide the desired ω_n; and (2) the gain K_D in the tachometer feedback path was selected to produce the desired ζ. The velocity error coefficient:

$$K_v = \frac{\omega_n}{2\zeta} = \frac{10}{2 \times 0.5} = 10 \text{ s}^{-1}$$

is therefore determined by the choices of ω_n and ζ.

Independent adjustment of a third performance parameter such as K_v becomes possible only if the servo loop structure is altered. The simplest way to modify $G(s)$ of Figure A.2 is to insert an appropriate network in the feedforward path. Such a network is called a *compensation network.* If $G_c(s)$ is the transfer function of the network, the modified open-loop transfer function $G_m(s)$ becomes

$$G_m(s) = G(s)G_c(s)$$

To illustrate an application of series compensation, suppose that we wish to reduce velocity lag of the antenna servo by increasing K_v by a factor of 10 while retaining essentially the original step response of Figure A.6. Recalling the definition of K_v, we see that a 10-fold increase in dc loop gain is required; thus the compensating network must produce a 20-dB drop in loop gain somewhere between zero frequency and the band of frequencies responsible for the essential features of the servo step response. If this can be accomplished, the existing step response will be essentially unaltered. A network that accomplishes this function is the lag network; its transfer function is

$$G_c(s) = \frac{sT_a + 1}{sT_b + 1}$$

in which $T_b > T_a$. This transfer function can be provided by the simple *RC* network of Figure A.9(a). If $T_b = 10T_a$, the Bode plot of Figure A.9(b) indicates that the network's transfer function amplitude drops 20 dB in the frequency decade between $\omega_b = 1/T_b$ and $\omega_a = 1/T_a$. To avoid appreciable alteration of the original step response, we must place this frequency decade well below the band of frequencies essential to that response. The existing $g(s)$ corner frequency is at

$$\omega = 2\zeta\omega_n = 10 \text{ rad/s}$$

Let us set $T_a = 4$ s, placing the upper corner frequency of $G_c(s)$ at $\omega_a = 1/T_a = 0.25$ rad/s. The lower corner frequency is then at $\omega_b = 1/T_b = 1/10T_a = 0.025$ rad/s. Placing ω_a and ω_b so low ensures that $G_c(j\omega)$ will have little impact on either amplitude or phase of the modified transfer function $G_m(j\omega)$ in the upper regions of the servo passband. Figure A.10 is a Bode plot for $G_m(j\omega)$ (dashed lines), with the original uncompensated $G(j\omega)$ superimposed (solid lines).

The compensated step response proves to be very close to the original response. Figure A.11 is a plot of error, $\theta_T - \Theta_A$, of the servo response to a unit

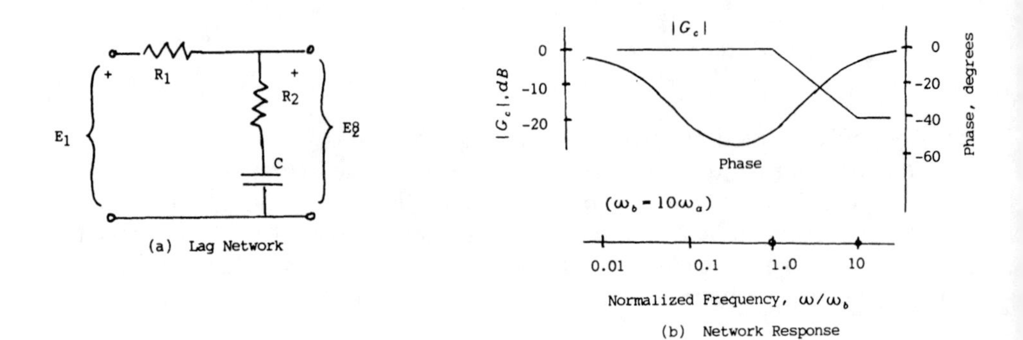

Figure A.9 Lag network response.

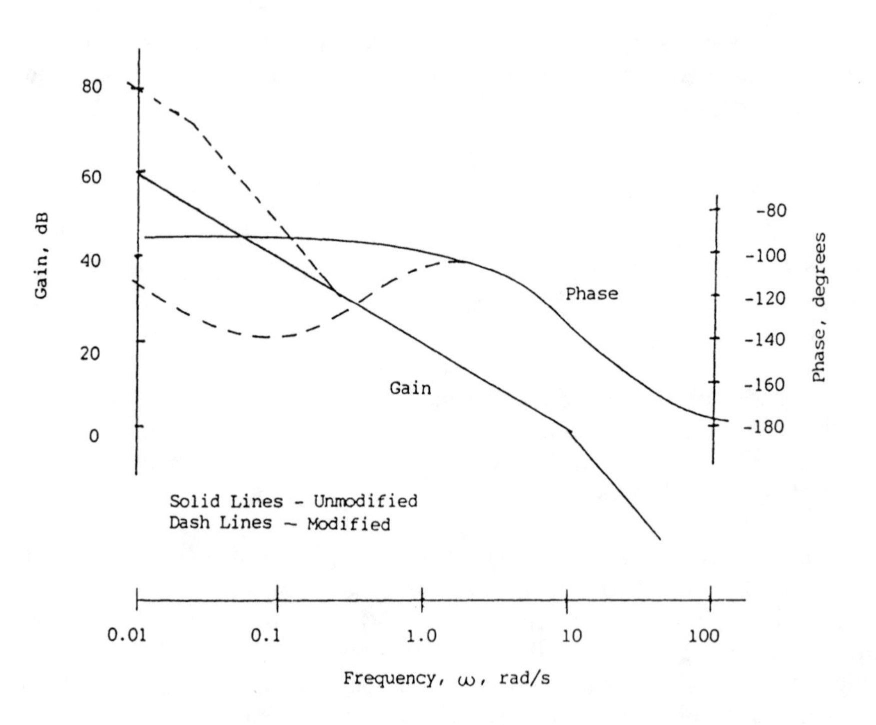

Figure A.10 Bode plots for modified and unmodified servo.

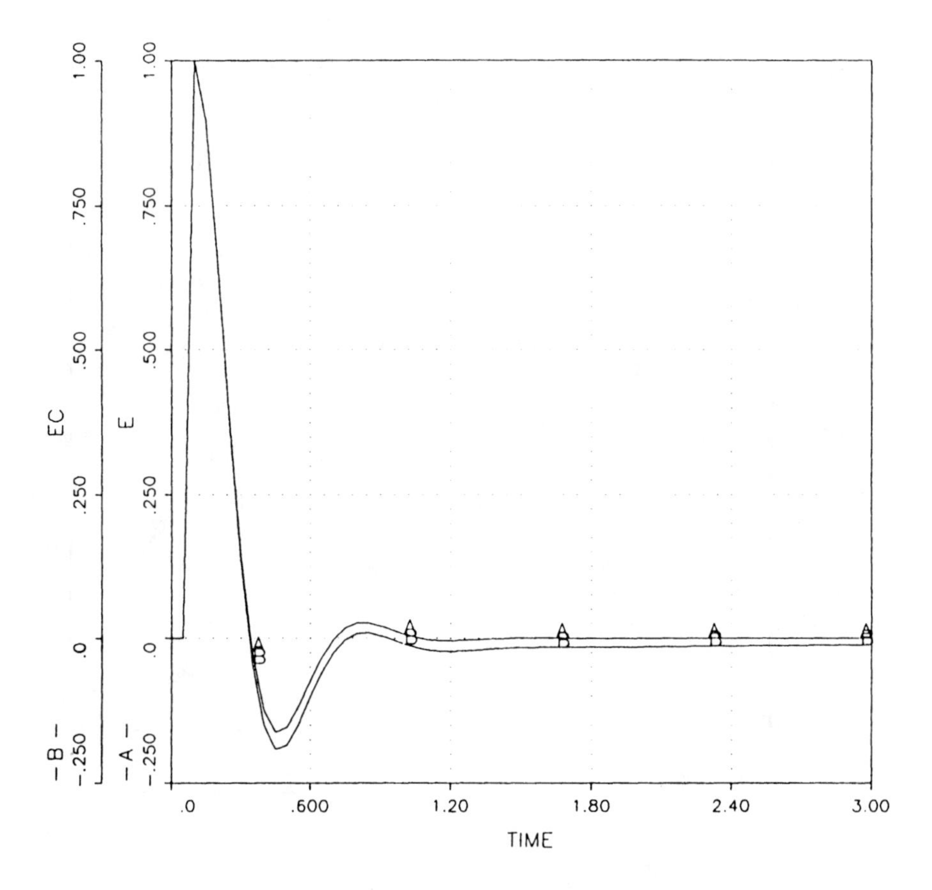

Figure A.11 Servo error in response to unit step (modified and unmodified servos).

step. The error ϵ_c of the compensated response departs from the error ϵ of the original response by only about 2% near $t = 1.5$ s. Figure A.12 shows the drastic change in velocity lag error we achieve by inserting $G_c(s)$. The input was a displacement ramp (a velocity step) of 5 rad/s. As expected, the lag error of the original servo settles out after the initial transient to the steady-state value:

$$\epsilon_{ss} = \frac{\dot{\theta}_t}{K_v} = \frac{5}{10} = 0.5 \text{ rad}$$

The transient error of the compensated servo follows that of the original servo for roughly the first second and then heads exponentially for the value:

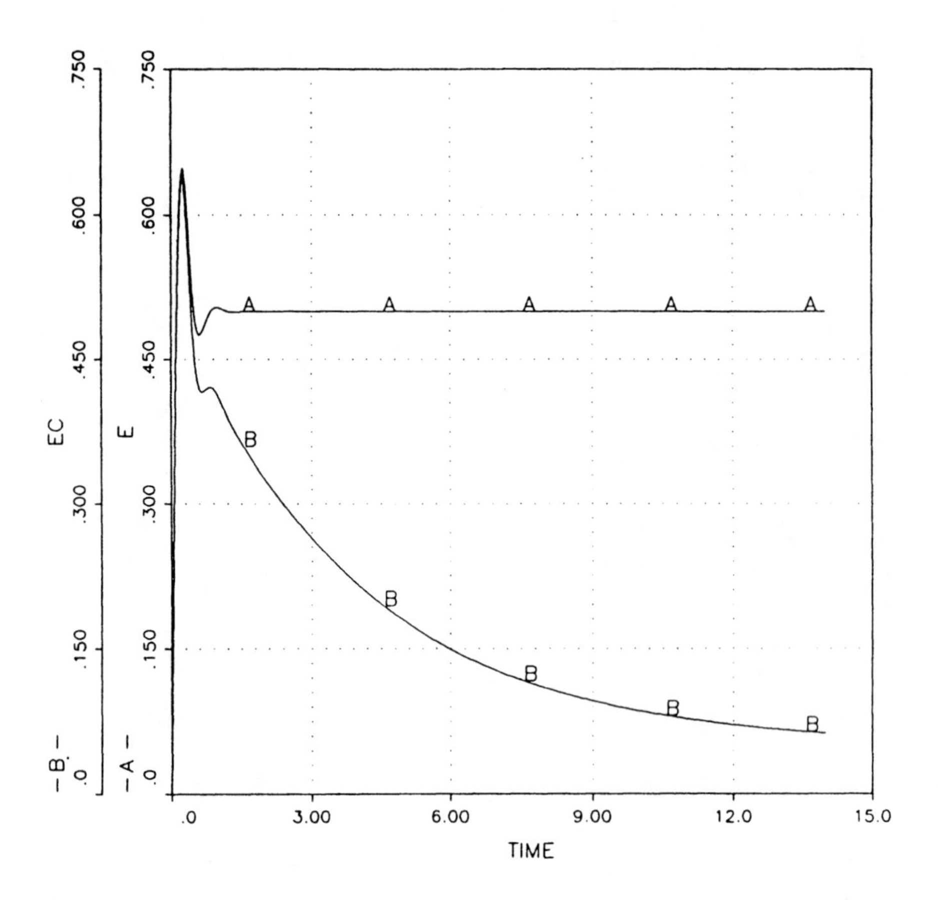

Figure A.12 Servo error in response to a rate input of 5.0 rad/s (modified and unmodified servos).

$$(\epsilon_c)_{ss} = \frac{\dot{\theta}_T}{(K_v)_c} = \frac{5}{100} = 0.05 \text{ rad}$$

Because of the very long time constants of the lag network, $(\epsilon_c)_{ss}$ is only down to about 0.1 rad at the end of 10 s. In a system that is subject to short-term input transients, such a long settling time would not be acceptable. It would be acceptable if one were interested only in improving position accuracy in a system that runs for long periods at essentially constant angular velocity.

This example provides one illustration of the limitations of compensation techniques. We could have accepted a smaller improvement in K_v and moved ω_b nearer to ω_a, thereby decreasing the $(\epsilon_c)_{ss}$ settling time. We could have retained the 20-dB improvement and moved the ω_b to ω_a decade higher in frequency, also decreasing settling time. But as the corner frequencies ω_b and ω_a are moved upward, the servo step response will be increasingly altered. One is forced to accept a compromise between transient response and settling time of $(\epsilon_c)_{ss}$.

REFERENCES

1. Bode, H.W., *Network Analysis and Feedback Amplifier Design,* Van Nostrand, New York, 1945.
2. James, H.M., N.B. Nichols, and R.S. Phillips, *Theory of Servomechanisms,* Vol. 25, M.I.T. Radiation Laboratory Series, McGraw-Hill, New York, 1947.
3. Greenwood, I.A., J.V. Holdam, Jr., and D. MacRae, Jr., *Electronic Instruments,* Vol. 21, M.I.T. Radiation Laboratory Series, McGraw-Hill, New York, 1948.
4. Chestnut, H., and R.W. Mayer, *Servomechanisms and Regulating System Design,* John Wiley, New York, 1951.
5. Gardner, M.F., and J.L. Barnes, *Transients in Linear Systems,* John Wiley, New York, 1942.
6. Souders, M., *The Engineer's Companion,* John Wiley, New York, 1966.
7. Truxal, J.G., *Control System Synthesis,* McGraw-Hill, New York, 1955.
8. Bruns, R.A., and R.M. Saunders, *Analysis of Feedback Control Systems,* McGraw-Hill, New York, 1955.

Appendix B
TARGET COORDINATE MEASUREMENT AND TARGET TRACKING

The final phase of an electronic warfare engagement takes place between a jammer and the fire control radar employed in the launch and steering of weapons. This appendix describes the target-tracking and target-measurement process inherent in the fire control radar.

Figure B.1 depicts a radar that performs its own search and target-designation functions. This diagram might describe the multifunction radar of an attack aircraft. Frequently, a separate search radar performs the initial detection and then hands over (designates) the target to the fire control radar. The block labeled $F(A, t, f_d, \theta, \phi)$ represents the combined antenna and receiver response function. The arguments of F are the target parameters:

A = amplitude of target return;
t = target range delay;
f_d = target doppler frequency;
θ, ϕ = target angle coordinates.

The value of A is of concern in that returns from potentially threatening targets should be high enough for reliable detection and accurate tracking, but often A is not one of the measured target parameters.[1] In a linear system, the amplitude of the response function will simply be proportional to A, so we can write

$$F(A, t, f_d, \theta, \phi) = AF_1(t, f_d, \theta, \phi)$$

where F_1 is the normalized response function (response to a target return of unit amplitude). The target's spatial position coordinates are always measured for use

[1]A simple calibration procedure permits A to be related to the radar cross section, σ_T, of the target. This parameter may, in some cases, enter into decisions to attack.

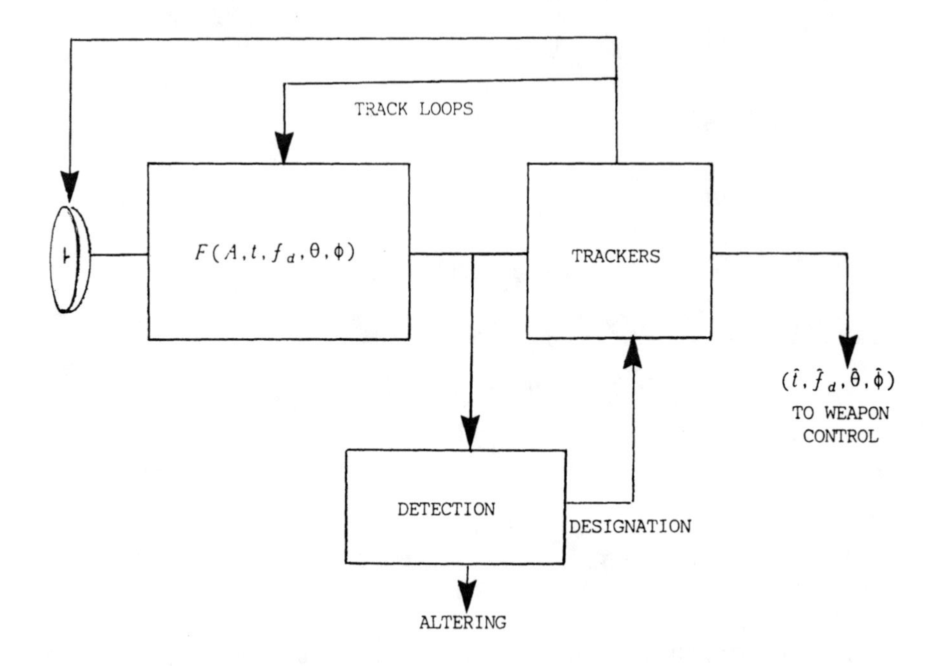

Figure B.1 Autonomous fire control radar.

in the solution of the weapon control problem. The range R is given in terms of range delay t by

$$R = \frac{ct}{2}$$

where c is the velocity of light. The target's doppler frequency f_d may or may not be measured. If measured, it reveals the target's radial velocity $\dot{R}$, given by

$$\dot{R} = \frac{\lambda f_d}{2}$$

where λ is the wavelength of the radar transmission. We shall see that often the concern with target doppler has little to do with the need to know target range rate, but, instead, arises out of a need to resolve and detect moving targets against a background of clutter, interference, and noise.

In Figure B.1, $\hat{t}$, $\hat{fd}$, $\hat{\theta}$, and $\hat{\phi}$ represent the estimates of the target coordinates as determined by the tracking loops. Although the description of the antenna-receiver response as a function of the four target coordinates is a valid one, it may be helpful to view F_1 as the product of the antenna response function F_a and the receiver response function F_b:

$$F_1(t, f_d, \theta, \phi) = F_a(\theta, \phi)F_b(t, f_d)$$

$F_a(\theta, \phi)$ describes the antenna beam pattern. Figure B.2 pictorially describes a response function $F(X_1, X_2)$ that could represent F_a or F_b. This function could be called a *resolution* function; it resolves or discriminates the target from its surroundings that may contain noise, clutter, and other targets. The spike labeled T in Figure B.2 represents a target with coordinates X_{1T}, X_{2T}. The target coordinates can be estimated (measured) by sliding the response function about in the X_1, X_2 plane until a peak response is found.

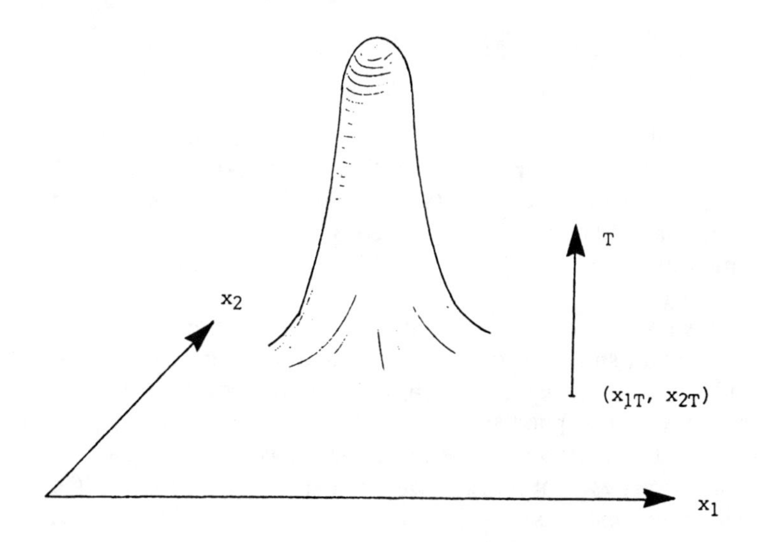

Figure B.2 Response function and target.

It is the job of the automatic coordinate tracker to keep the response function centered on the target. It is easy to visualize the manual tracking process. One would apply a dither (a small back-and-forth motion) to the response function along one coordinate axis to locate the peak response, and then repeat the process for the other axis. Some automatic trackers utilize this scheme. Conical scan, employed for angle tracking, can be viewed as a simultaneous dither of both axes. The one-dimensional dither process is a way of finding the derivative of the response function in one coordinate. The sign (polarity) of the derivative points the direction to the peak, where the derivative passes through zero and changes sign. Thus, the derivative has two of the essential properties of an error correction signal for driving the automatic track loop: (1) Its sign tells the direction of the correction that must be made; and (2) it goes to zero when no further correction is needed.

As Barton shows [1, Chapter 8], the derivative of the response function in one of the coordinates can be approximated by the difference, $\Delta(x)$, of a pair of overlapping response functions:

$$\Delta(x) = f_1(x) - f_2(x)$$

The functions f_1 and f_2 are identical response functions, offset by an amount δx on either side of the measurement axis x_0. Clearly, the difference function:

$$\frac{\Delta(x)}{2\delta x} = \frac{f(x_0 + \delta x) - f(x_0 - \delta x)}{2\delta x}$$

does approach the derivative at $x = x_0$ as δx becomes small. Actually, δx may be a significant fraction of the nominal width of the response function.

Figure B.3(b) shows $\Delta(x)$ obtained from the two offset functions of Figure B.3(a). Note that $\Delta(x)$ has the two desired properties of an error correction signal already mentioned. It has a third desirable property in that, for small departures from the measurement point x_0, it is a relatively linear function of x. This means that if $\Delta(x)$ is used for error sensing, it contributes no nonlinearities to the control system (at least for small errors). We can generate the response functions f_1 and f_2 sequentially in time (sequential lobing) or simultaneously (simultaneous lobing). The former has been applied as lobe switching for angle tracking in radar systems. In fact, the conical scan technique can be viewed as sequential lobing, although the antenna beam moves continuously rather than in discrete steps. The monopulse angle-tracking technique is a much-used simultaneous lobing scheme.

The function $\Delta(x)$ has one shortcoming as an error correction signal; its amplitude is proportional to the amplitude of the target return. If $\Delta(x)$ were used directly as the error correction signal, the gain of the tracking loop would be proportional to the target return amplitude. This would be unacceptable. A tracking servomechanism design must be based on a fixed value of loop gain, for this will govern both transient behavior and steady-state tracking errors.[2] We can remove

[2]See Appendix A.

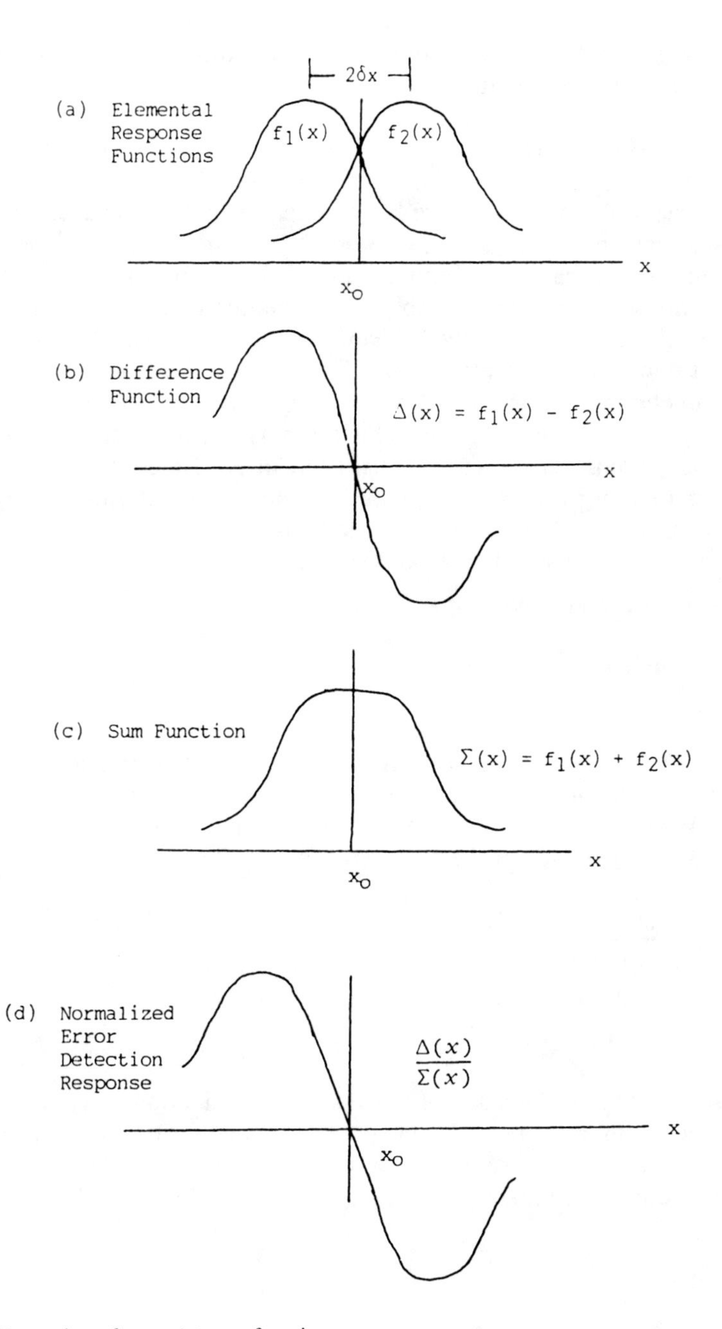

Figure B.3 Formation of error detector function.

the dependence on target return amplitude by dividing $\Delta(x)$ by $\Sigma(x)$, the sum of the two offset response functions:

$$\Sigma(x) = f_1(x) + f_2(x)$$

We plot the sum function in Figure B.3(c). Because both Δ and Σ are proportional to the target return amplitude, the quotient Δ/Σ is independent of the amplitude. The division process can be performed numerically in a digital system. In an analog system the division is effected by applying automatic gain control, in which the gain control signal is derived from the Σ channel [2, Chapter 21]. The division of Δ by Σ is called *normalizing* the error signal. This process yields an additional benefit; for most functions $f(x)$, the normalized error signal $\Delta(x)/\Sigma(x)$ is linear over a wider range of x on either side of x_0 than is the nonnormalized function $\Delta(x)$.

The slope of the error detector characteristic is of interest for two reasons: (1) It becomes a factor in the tracking loop gain; and (2) it determines the sensitivity of the system to disturbances such as system noise. Let us derive the error detector slope for a specific form of the response function $f(x)$.

Example: Let the elemental response function be

$$f(x) = \frac{\sin(ax)}{ax}$$

where $a = 2.78/x_3$ and x_3 is the 3-dB width of $f(x)$. It is immaterial whether we view $f(x)$ as an antenna voltage pattern, a filter amplitude response, or a time-gating function. We shall choose $x_0 = 0$ as the measurement point, and offset $f_1(x)$ and $f_2(x)$ by δx on either side of the point $x = 0$:

$$f_1(x) = \frac{\sin[a(x + \delta x)]}{a(x + \delta x)}$$

$$f_2(x) = \frac{\sin[a(x - \delta x)]}{a(x - \delta x)}$$

We want to compute the slope of the normalized function $\Delta(x)/\Sigma(x)$ at $x = 0$, which is the equilibrium point (stable tracking point) for the x-coordinate tracker. To simplify the notation, we use Δ', Σ', f_1', and f_2' to represent the first derivatives of the functions Δ, Σ, f_1, and f_2. The slope expression becomes

$$\text{slope} = \frac{\mathrm{d}}{\mathrm{d}x}\left(\frac{\Delta}{\Sigma}\right) = \frac{\Sigma\Delta' - \Delta\Sigma'}{\Sigma^2}$$

The slope is to be evaluated at $x = 0$, where, because of the even symmetry of $f(x)$, we can write

$$\text{slope}\ |_{x=0} = \frac{f_1'(0)}{f_1(0)}$$

The numerator and denominator expressions are

$$f_1'(0) = \frac{a^2\delta x\ \cos a\delta x) - a\ \sin(a\delta x)}{a^2\delta x^2}$$

and

$$f_1(0) = \frac{\sin(a\delta x)}{a\delta x}$$

The slope expression reduces to

$$\text{slope}\ |_{x=0} = \left[a\,\frac{\cos(a\delta x)}{\sin(a\delta x)} - \frac{1}{\delta x} \right]$$

Let us evaluate the slope for $\delta x = 0.3x_3$:

$$\text{slope}\ |_{x=0} = \left[\frac{2.78\ \cos(2.78 \times 0.3)}{x_3\ \sin(2.78 \times 0.3)} - \frac{1}{0.3x_3} \right]$$

$$\text{slope}\ |_{x=0} = \frac{0.81}{x_3}$$

Barton [1, Chapter 8] defines the following normalized error slope function:

$$k_0 = \frac{\mathrm{d}(\Delta/\Sigma)}{\mathrm{d}(x/x_3)}\ |_{x=0}$$

This slope function expresses the ratio of the incremental Δ channel output (as a fraction of the Σ channel output) to the incremental error measured in units of resolution "lobe width" x_3. For our example, this slope constant becomes $k_0 = -0.81$.

Relation of Error Slope to Loop Gain

We now consider how the slope function k_0 is related to tracking loop gain. Figure B.4 shows a closed-loop tracking servomechanism. The target coordinate is x, and the tracker output (the measured value of the target coordinate) is $\hat{x}$. The tracking error is

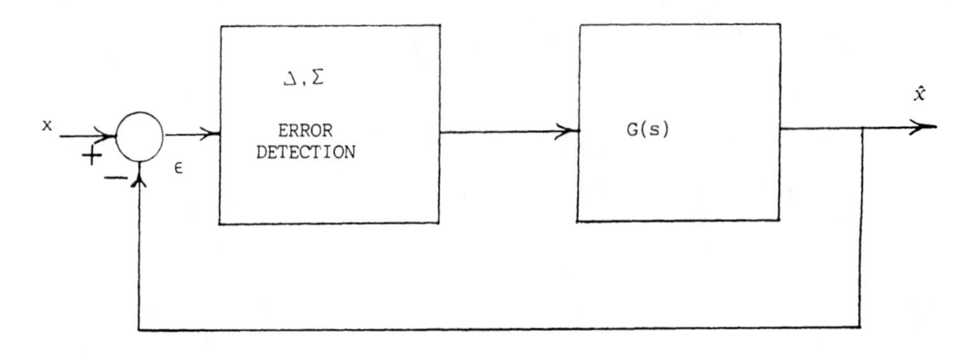

Figure B.4 Radar Coordinate Tracker.

$$\epsilon = x - \hat{x}$$

A Δ, Σ error detector, as discussed earlier, develops an error voltage:

$$V_\epsilon = \frac{\Delta(\epsilon)}{\Sigma(\epsilon)}$$

In terms of k_0, the error voltage is

$$V_\epsilon = k_0 \left(\frac{\epsilon}{x_3} \right)$$

so the error detector can be said to have a detector constant:

$$k_\epsilon = \frac{k_0}{x_3}$$

This detector constant becomes a factor in the loop gain expression. Thus for the loop of Figure B.4, the loop gain at frequency ω_1 is

$$\text{gain} \,\big|_{\omega=\omega_1} = k_\epsilon G(j\omega_1)$$

where $G(s)$ represents the open-loop transfer function[3] of the portion of the system beyond the error detector.

[3]See Appendix A for a discussion of transfer functions.

Relation of Tracking Error to Error Slope

Barton [1, Chapter 8] discusses the effects of random and deterministic interferences on the accuracy of tracking. In the foregoing discussion, $\Delta(x)$ and $\Sigma(x)$ were treated as real functions. For sinusoidal signals (that is, signals containing a carrier, either at RF or IF) the outputs of the Δ and Σ channels can be considered to be phasors. In the two channels before the detection point, the channel phase shifts have been adjusted so that for a target on one side of the measurement axis (say $x > 0$) the Σ and Δ channel signals are in phase, and for a target on the other side of the measurement axis ($x < 0$) the Σ and Δ channel signals are 180° out of phase. This is essential to ensure the preservation of the sense (polarity) of the error; synchronous detection (phase detection) of the Δ channel output, using the Σ channel as a phase reference, produces a detector output of polarity corresponding to the sense of the error. An interfering signal arriving from outside the tracking system arrives with an amplitude V_i and an x-coordinate x_i. If V_i is much smaller than the amplitude V_t of the target being tracked, V_t mainly determines the Σ channel output that serves as the basis for the normalized (Δ/Σ) error voltage output. Therefore, the interference produces an interference contribution, Δ_i/Σ_t, at the error detector output, where Δ_i is the difference channel voltage due to the interference and Σ_t is the sum channel voltage due mainly to the target. The tracking loop seeks to null the error detector output, so it drives the target off axis by an amount x_ϵ such that the target-induced contribution:

$$\frac{x_\epsilon}{x_3} = \frac{1}{k_0}\frac{\Delta_t}{\Sigma_t}$$

just balances the interference contribution at the detector output. In this expression, Δ_t is the difference channel target signal. The result is

$$\frac{\Delta_i}{\Sigma_t} + \frac{\Delta_t}{\Sigma_t} = 0 = \frac{\Delta_i}{\Sigma_t} + \frac{k_0}{x_3}x_\epsilon$$

or

$$\frac{x_\epsilon}{x_3} = \frac{-1}{k_0}\frac{\Delta_i}{\Sigma_t}$$

This is the error (in units of lobe width x_3) caused by the interfering signal. If the interfering signal's coordinate, x_i, is within the linear range of the error detector:

$$\frac{\Delta_i}{\Sigma_i} = k_0\frac{x_i}{x_3}$$

and the resulting error becomes

$$\frac{x_\epsilon}{x_3} = -\frac{\Sigma_i}{\Sigma_t}\frac{x_i}{x_3}$$

However, because both target and interference fall not far from the center of the sum pattern:

$$\frac{\Sigma_i}{\Sigma_t} \approx \frac{V_i}{V_t}$$

with the result that the tracking error is approximately

$$x_\epsilon = -\frac{V_i}{V_t}x_i$$

This expression is valid only for $V_i \ll V_t$ and x_i near x_t.

When the interference is random circuit noise, x_i has no meaning because the interfering voltage v_i is internal to the receiver that feeds the error detector. The same process occurs in which tracking errors are produced that offset the interference contribution at the error detector output, but now both the amplitude and phase of the errors are random, so it is no longer meaningful to carry along the minus sign of the foregoing equations. If the noise voltage of the difference channel, $V_{\Delta n}$, were always in phase or 180° out of phase with the normalizing voltage $V_{\Sigma t}$, we could write, following our previous argument:

$$\frac{x_\epsilon}{x_3} = \frac{1}{k_0}\frac{V_{\Delta n}}{V_{\Sigma t}}$$

understanding that $V_{\Sigma t} \gg V_{\Sigma n}$, so the target dominates the sum channel. This equation is not valid because $V_{\Delta n}$ is a random noise waveform, and, considered as a phasor, only half the noise power is in phase with the sum channel reference $V_{\Sigma t}$. Therefore only the rms component $V_{\Delta n}/\sqrt{2}$ is effective in producing offsetting track errors. (The quadrature component is rejected by the synchronous detection process.) Therefore we must write for the rms tracking error:

$$\frac{x_\epsilon}{x_3} = \frac{1}{k_0}\frac{V_{\Delta n}}{\sqrt{2}V_{\Sigma t}}$$

This is, in effect, Barton's [1, Chapter 8] equation (8.1.9), provided that x_ϵ is replaced by σ_x, representing the rms tracking error, and $V_{\Delta n}/V_{\Sigma t}$ is replaced by

$\sqrt{I_\Delta/S}$, where S is the signal power in the sum channel and I_Δ is the interfering noise power in the difference channel.

Summary

We began by defining the multidimensional (multicoordinate) response function of a tracking radar as a function that peaks at the point corresponding to the target coordinates. Next, we discussed the independent tracking loops, one for each of the target coordinates. The desired peak response occurs only if all the loops are locked to the target. Achieving the peak response is important not only for its own sake, but also because the properly peaked response discriminates against extraneous sources of interference. The range tracker cannot properly track the target if the antenna beam slips off the target, losing target signal power and picking up nearby clutter. Likewise, the angle tracker cannot properly function unless the range gate selects the target and rejects the noise and clutter present in all the other range cells. As noted earlier, doppler tracking is employed in some radars, but in many it is not. In those radars that do not employ doppler tracking, the multidimensional response function can be said to be unpeaked in the doppler dimension; it has a ridge of essentially uniform height running along the doppler axis. In fact, the signal needs to be properly designed to make a peaked doppler response (doppler resolution) possible. Mainly, it must have adequate total duration. A long CW signal is ideal in this respect, but it provides essentially no range resolution. Long individual pulses are rarely long enough to yield useful doppler measurements. The alternative is to employ long coherent pulse trains, and this approach is taken in radars that must find and track moving targets against strong clutter backgrounds, as is true of airborne radars that must look down to find and track low-flying aircraft. Such radars are called *pulsed doppler radars.*

REFERENCES

1. Barton, D.K., *Modern Radar Systems,* Artech House, Norwood, MA, 1988.
2. Skolnik, M.I., ed., *Radar Handbook,* McGraw-Hill, New York, 1970.

Appendix C
RADAR-JAMMING SIMULATION

Background

In jamming problems, high *J/S* often drives the radar receiver into nonlinear modes of operation, rendering analysis difficult or impossible. A reasonable approach then is to resort to simulation. One writer [1, Chapter 3] states, "System simulation is the technique of solving problems by the observation of the performance over time, of a dynamic model of the system."

In the past, analog computers served as the primary means for the dynamic modeling of electromechanical systems. A major virtue of analog simulation was that programming consisted of interconnecting the appropriate building blocks according to the known cause-and-effect relations governing the physical system being simulated. Thus the simulation diagram (which served as a program) strongly resembled a block diagram of the physical system, and the monitoring of voltages at block input-output points gave the feeling of watching the physical system in operation. Another virtue of analog simulation is that once simulation is started (after setting proper initial conditions), integration of the differential equations of the system proceeds simultaneously and continuously throughout the system with no demands for programming beyond making the proper connections. That is, integration becomes a parallel computing operation with all the integrators automatically synchronized simply because the independent variable is time.

Due to the following shortcomings, analog computers have been almost completely replaced by digital methods:

1. *Limited dynamic range.* Voltages representing system variables run into saturation or become too low for accurate measurement. Rescaling is then necessary.
2. *Limited accuracy of components.* Analog computing elements might typically have accuracies of one part in 10^3 as a fraction of full scale.
3. *Complexity of nonlinear operations.* Although the linear elements (operational amplifiers, integrators, *et cetera*) were simple and effective, nonlinear operations (multiplication, trigonometric function generation, *et cetera*) required special-purpose elements that were relatively expensive and were

often slow compared with the linear elements (e.g., servo resolvers, servo multipliers, and x-y plotters).

4. *Expense of reconfiguring.* Although a general-purpose digital computer can quickly switch from one prepared program to another, this was not true of most analog facilities.

Digital Simulation

Digital methods have prevailed over analog simulation methods because

1. Digital computation became very inexpensive.
2. With floating-point representation of the values of system variables, the dynamic range problem was eliminated.
3. Drift effects are nonexistent, no matter how long the computation requires.
4. Essentially any linear or nonlinear operation can be performed digitally to whatever accuracy is required.
5. An unlimited variety of simulation programs (as well as other programs) can be stored compactly on various media. Switching from one program to another consumes negligible time, in comparison with the time required for reprogramming an analog computer.

This is not to imply that digital simulation is free of problems. Unless considerable thought is given to program structure, a simulation program could become a jumble of equations bearing little apparent relation to the structure of the system being simulated. It is important that the program be structured in conformity with the cause-and-effect relations of clearly identifiable functional blocks of the physical system. Program blocks can then be debugged much as physical subsystems are tested. Moreover, modification of the model is facilitated, for it becomes possible to bypass or replace individual blocks or combinations of blocks to create a new system model.

Digital simulation has an inherent feature that complicates this neat packaging of program blocks. A program flow diagram differs from a block diagram of a physical system in that the former defines a *sequence* of computing operations, whereas the latter defines a *structure* of elements and their interconnections. In the analog computer the synchronization of the integration process is natural and automatic. In the digital simulation this is not the case, for usually one central processor must sequentially service all the integrations of the system model.[1] This synchronization must be forced by the overall program, so if program blocks are replaced or moved about this sequencing operation cannot be overlooked, for the value of

[1]This sequential operation can be replaced by simultaneous (parallel) operation if multiple processors are available. All integrations employing a common integration step size can then be grouped together, with the integrators executing their finite integration steps in synchronism.

the independent variable must be kept in synchronism throughout the simulation model.

Numerical Integration of Differential Equations

Various numerical integration techniques are described in books [1–3] on simulation and numerical mathematical analysis, sometimes under the heading of "numerical quadrature." To illustrate the application of numerical integration to simulation, let us consider the angle-tracking example of Appendix A. The open-loop transfer function was

$$G(s) = \frac{K}{s(s\tau + 1)}$$

and the closed-loop transfer function was

$$\frac{Y(s)}{X(s)} = \frac{G(s)}{1 + G(s)} = \frac{K}{s(s\tau + 1) + K} \tag{C.1}$$

$x(t)$ and $Y(t)$ were, respectively, the angular position of the target and the angular position of the tracking antenna. The second-order differential equation corresponding to (C.1) is

$$\tau\ddot{y} + \dot{y} + Ky = Kx \tag{C.2}$$

Equation (C.2), solved for the highest derivative $\ddot{y}$, yields

$$\ddot{y} = f(x, y, \dot{y}) \tag{C.3}$$

where

$$f(x, y, \dot{y}) = \frac{1}{\tau}[Kx - Ky - \dot{y}] \tag{C.4}$$

We can replace the original second-order equation by the following pair of first-order equations:

$$\begin{aligned} \frac{\mathrm{d}\dot{y}}{\mathrm{d}t} &= f(x, y, \dot{y}) \\ \frac{\mathrm{d}y}{\mathrm{d}t} &= \dot{y} \end{aligned} \tag{C.5}$$

These equations yield, when integrated,

$$\dot{y} = \int_0^t f(x, y, \dot{y})\, \mathrm{d}t$$
$$y = \int_0^t \dot{y}\, \mathrm{d}t \qquad \text{(C.6)}$$

To integrate these equations, we select a finite small increment Δt of the independent variable, and, knowing the value of the integral at a particular instant of time, we approximate its value at a time Δt later by a suitable numerical integration algorithm. Thus, we proceed to compute the values of the integral at the times $t = 0$, Δt, $2\ \Delta t$, $3\ \Delta t$, ... The simplest integration algorithm is rectangular integration (sometimes called Euler's method). This method applied to (C.6) yields

$$\dot{y}_{n+1} = \dot{y}_n + (\Delta t) f_n \qquad \text{(C.7)}$$
$$y_{n+1} = y_n + (\Delta t)\dot{y}_n$$

where $y_n = y(n\ \Delta t)$ and $f_n = f[x(n\ \Delta t), y(n\ \Delta t), \dot{y}(n\ \Delta t)]$. In this simple algorithm, the values of y and its first derivative at any instant are computed from their values plus the value of f at the end of the preceding step. Other, more complex, algorithms can achieve acceptable accuracy with larger Δt increments than are permissible with rectangular integration.

This example is sufficient to indicate the requirement for maintaining synchronization of integration steps throughout the model of the physical system. The input, $x(t)$, that enters into $f(x, y, \dot{y})$ arises from motion of a radar target (e.g., an aircraft) for which the dynamic behavior may be described by another set of differential equations, and these equations must be integrated in synchronism with the equations we have been considering. The step-by-step process is fairly simple for the integration algorithm we discussed, but it becomes more complex with algorithms that use not only variable values from the immediately preceding step but perhaps from many preceding steps. Some algorithms even involve predicted values of variables (predicted for $t = (n + 1)\ \Delta t$ and even further ahead). Moreover, some algorithms execute an iteration procedure at a given $t = n\ \Delta t$ to improve the integration accuracy before they move to the next step.

All this bookkeeping to maintain synchronism of the integration process is burdensome to the analyst whose real interest lies not in numerical techniques but in the dynamics of the physical system. This burden has been removed through the development of special simulation languages.

Simulation Languages

We noted that analog simulation provided great convenience for working with the model, such as making design changes, varying parameter values, and measuring system responses. Digital simulation languages have been developed to give the analyst essentially the same convenience and freedom from concern with the internal computation process.

In these simulation languages, special function subroutines are provided for modeling linear dynamic system blocks that the engineer commonly describes in terms of transfer functions. Special nonlinear function blocks are also provided. In the automatic compilation process (conversion of source program into machine-executable instructions) the calls to the special simulation routines result in a translation into another high-level general-purpose language (e.g., FORTRAN). The translated instructions are interleaved in proper order[2] with any instructions written in the general-purpose language. Then the entire program is compiled. This program contains the provisions for inserting the proper initial conditions for the integrals. These automatic programming aids permit the analyst to put together a model in very much the same way as interconnecting analog computer blocks. We used SYSL (System Simulation Language) [4] for the radar-jamming simulation for this book.

Levels of Simulation Detail

In a digital simulation, all the variables that are continuous functions of time[3] must be represented by sequences of discrete values, generally the values of the variable at equally spaced time increments. One can imagine a radar-ECM simulation in which the time increment Δt is so small that even the individual cycles of IF or RF waveforms are faithfully represented by the discrete samples. Not only would the simulation of such fine detail consume great amounts of memory, but it would also consume a great deal of computing time. Such detail is usually neither necessary nor useful. For instance, the dynamics of a target tracker are normally characterized by time constants of the order of seconds. Within one time constant there would be millions of cycles of the IF waveform and probably billions of RF cycles. The time increments need to be fine enough to provide perhaps tens of samples (or even hundreds) per system time constant, but not millions or billions. The RF or IF carrier is important to the system only in that it serves as the vehicle for the envelope[4] of the waveform. Thus, in the simulation of a pulse waveform, we generally need only a single pair of samples (the amplitude of the pulse and the phase of the carrier relative to some reference) to represent the pulse. Moreover, it is not

[2]The executable (object) program is executed sequentially. Therefore, instructions must be placed in an order that provides for computation of the value of a given variable before this value is needed for a subsequent computation of another variable's value.

[3]In our dynamic system simulations, time is the independent variable. Other kinds of simulation problems can involve other independent variables.

[4]In a coherent system, the "complex" envelope (that is, both the amplitude and phase of the carrier) is of concern.

always necessary to carry the simulation detail to the point of describing the amplitude and phase of each pulse. For instance, in a target tracker there will generally be many pulse returns within the span of one time constant of the tracking servo. The tracking loop then smooths or averages over many pulse samples. The simulation result based on a continuous representation of target return (as might be obtained from a CW radar) will then be little different from the result of a simulation in which the individual pulse samples are modeled.

For simplicity, we used this continuous representation of target returns in many simulations described in this book. This simplification allowed us to choose the integration step size Δt without regard to actual pulse timing. We might picture this simplification as the result of passing the discrete pulse samples through a boxcar (sample-and-hold) circuit, followed by a smoothing operation. This simplified model is not adequate if the PRI is so low that there are not many pulse samples per system time constant. In some cases it will be possible to make Δt equal to the radar PRI. In others we may have to choose Δt much smaller than the radar PRI to maintain adequate accuracy in the integration process. It may then be convenient to make Δt a submultiple of the PRI.

REFERENCES

1. Gordon, G., *System Simulation,* 2nd Ed., Prentice-Hall, Englewood Cliffs, NJ, 1978.
2. Scarborough, J.B., *Numerical Mathematical Analysis,* The Johns Hopkins Press, Baltimore, 1950.
3. Kochenburger, R.J. *Computer Simulation of Dynamic Systems,* Prentice-Hall, Englewood Cliffs, NJ, 1972.
4. SYSL, System Simulation Language User's Guide, E^2 Consulting, Poway, CA.

INDEX

THE AUTHORS

Robert Lothes earned his BEE and MSEE at Ohio State University. As Staff Consulting Engineer, Defense Electronics Engineering Division at Syracuse Research Corporation for over 20 years, he was involved in numerous projects, some of which involved ECM for ballistic missile defense, beacon detection, and acquisition and tracking performance, with and without jamming signals. He holds two patents and is now an independent consultant.

Michael Szymanski earned his BS in Mathematics and his BS in Engineering Science at Renesselaer Polytechnic Institute. He spent three years as a USAF Lieutenant and Government Project Engineer at the Rome Air Development Center. Since 1977, he has been engaged in intelligence and electronic warfare efforts and works as a Senior Research Engineer for Research Associates of Syracuse. He is the coauthor of a book on pulse train analysis.

Richard Wiley earned his BS and MSEE at Carnegie-Mellon Univesity and his PhD at Syracuse University. His recent work as Vice President-Chief Scientist of Research Associates of Syracuse, Inc., includes national intelligence and electronic warfare efforts and the analysis of radar developments and trends. Dr. Wiley has authored three books, including *Electronic Intelligence: The Interception of Radar Signals* (Artech House, 1985), and *Electronic Intelligence: The Analysis of Radar Signals* (Artech House, 1982). He is a Fellow of the IEEE.

The Artech House Radar Library

David K. Barton, *Series Editor*

Active Radar Electronic Countermeasures by Edward J. Chrzanowski

Airborne Pulsed Doppler Radar by Guy V. Morris

AIRCOVER: Airborne Radar Vertical Coverage Calculation Software and User's Manual by William A. Skillman

Analog Automatic Control Loops in Radar and EW by Richard S. Hughes

Aspects of Modern Radar, by Eli Brookner, *et al.*

Aspects of Radar Signal Processing by Bernard Lewis, Frank Kretschmer, and Wesley Shelton

Detectability of Spread-Spectrum Signals by Robin A. Dillard and George M. Dillard

Electronic Homing Systems by M.V. Maksimov and G.I. Gorgonov

Electronic Intelligence: The Analysis of Radar Signals by Richard G. Wiley

Electronic Intelligence: The Interception of Radar Signals by Richard G. Wiley

EREPS: Engineer's Refractive Effects Prediction System Software and User's Manual, developed by NOSC

Handbook of Radar Measurement by David K. Barton and Harold R. Ward

High Resolution Radar by Donald R. Wehner

High Resolution Radar Imaging by Dean L. Mensa

Interference Suppression Techniques for Microwave Antennas and Transmitters by Ernest R. Freeman

Introduction to Electronic Warfare by D. Curtis Schleher

Introduction to Sensor Systems by S.A. Hovanessian

Logarithmic Amplification by Richard Smith Hughes

Modern Radar System Analysis by David K. Barton

Monopulse Principles and Techniques by Samuel M. Sherman

Monopulse Radar by A.I. Leonov and K.I. Fomichev

Multifunction Array Radar Design by Dale R. Billetter

Multisensor Data Fusion by Edward L. Waltz and James Llinas

Multiple-Target Tracking with Radar Applications by Samuel S. Blackman

Multitarget-Multisensor Tracking: Advanced Applications, Yaakov Bar-Shalom, ed.

Over-the-Horizon Radar by A.A. Kolosov, et al.

Principles and Applications of Millimeter-Wave Radar, Charles E. Brown and Nicholas C. Currie, eds.

Principles of Modern Radar Systems by Michael C. Stevens

Pulse Train Analysis Using Personal Computers by Richard G. Wiley and Michael B. Szymanski

Radar and the Atmosphere by Alfred J. Bogush, Jr.

Radar Anti-Jamming Techniques by M.V. Maksimov, *et al.*

Radar Cross Section by Eugene F. Knott, *et al.*

Radar Detection by J.V. DiFranco and W.L. Rubin

Radar Propagation at Low Altitudes by M.L. Meeks

Radar Range-Performance Analysis by Lamont V. Blake

Radar Reflectivity Measurement: Techniques and Applications, Nicholas C. Currie, ed.

Radar Reflectivity of Land and Sea by Maurice W. Long

Radar System Design and Analysis by S.A. Hovanessian

Radar Technology, Eli Brookner, ed.

Radar Vulnerability to Jamming by Robert N. Lothes, Michael B. Szymanski and Richard G. Wiley

Receiving Systems Design by Stephen J. Erst

RGCALC: Radar Range Detection Software and User's Manual by John E. Fielding and Gary D. Reynolds

SACALC: Signal Analysis Software and User's Guide by William T. Hardy

Secondary Surveillance Radar by Michael C. Stevens

SIGCLUT: Surface and Volumetric Clutter-to-Noise, Jammer and Target Signal-to-Noise Radar Calculation Software and User's Manual by William A. Skillman

Signal Theory and Random Processes by Harry Urkowitz

Solid-State Radar Transmitters by Edward D. Ostroff, *et al.*

Space-Based Radar Handbook, Leopold J. Cantafio, ed.

Statistical Theory of Extended Radar Targets by R.V. Ostrovityanov and F.A. Basalov

Techniques of Radar Reflectivity Measurement by Nicholas C. Currie

The Scattering of Electromagnetic Waves from Rough Surfaces by Petr Beckmann and Andre Spizzichino

VCCALC: Vertical Coverage Plotting Software and User's Manual by John E. Fielding and Gary D. Reynolds

Printed in the United States
124132LV00009B/24/A

9 780890 063880